# Collins

# Geography Fieldwork & Skills
## AS/A level Geography

## Barnaby Lenon
Former Head of Geography, Eton College

## Paul Cleves
Former Head of Geography, Eton College and St Paul's Girls' School

Published by Collins
An imprint of HarperCollins*Publishers* Ltd
1 London Bridge Street
London SE1 9GF

www.harpercollins.co.uk

HarperCollins*Publishers*
Macken House, 39/40 Mayor Street Upper, Dublin 1,
D01 C9W8, Ireland

**Browse the complete Collins catalogue at
www.collins.co.uk**

Text © Barnaby Lenon and Paul Cleves 1994, 2001, 2015
© HarperCollins*Publishers* Limited 2015
Previously published as *Fieldwork Techniques and Projects in Geography* 1994. Second edition 2001.

10 9 8 7

ISBN 978-0-00-759282-1

The authors assert their moral rights to be identified as the authors of this work.

All rights reserved. No part of this publication may be reproduced, stored in a retrieval system, or transmitted in any form or by any means, electronic, mechanical, photocopying, recording or otherwise, without the prior written permission of the Publisher or a licence permitting restricted copying in the United Kingdom issued by the Copyright Licensing Agency Ltd., 90 Tottenham Court Road, London W1T 4LP.

British Library Cataloguing in Publication Data
A Catalogue record for this publication is available from the British Library

Cover design: Angela English
Original design: Jacky Wedgewood and Sally Boothroyd
Original artwork: Jerry Fowler
Original picture research: Caroline Thompson

Printed and Bound in the UK by Ashford Colour Press Ltd.

MIX
Paper | Supporting responsible forestry
FSC™ C007454

## Acknowledgements

The author and publisher wish to thank the following for permission to use copyright material.

Fig. 1.1, Naish, M, Rawling, E and Hart, C (1987), *Geography 16–19: The contribution of a curriculum project to 16–19 education*, Longman, © SCAA; fig. 3.1, Goudie, A, Cooke, R and Evans, R (1970), 'Experimental investigation of rock weathering by salts', *Area*; fig. 3.3, Viles, H and Cooke, R (1991), 'Crumbling walls', *Geography Review*; fig. 3.12, Daniel, J (1971), 'Channel movement of meandering Indiana streams', United States Geographical Society; fig. 3.13, Hooke, J (1977), *River channel changes*, Wiley; fig. 3.18, Gregory, K and Cullingford, R (1974), 'Lateral variations in pebble shape in north-west Yorkshire', *Sedimentary Geology*; fig. 3.19, Powers, M (1953), 'A new roundness scale for sedimentary particles', *Journal of Sedimentary Petrology*; fig. 3.23, Goudie, A and Gardner, R (1985), *Discovering landscape in England and Wales*, Allen & Unwin; table 4.1, Chow, V T (1964), *Handbook of applied hydrology*, McGraw-Hill, New York, reproduced with permission from McGraw-Hill; fig. 4.9, Petts, G (1983), *Rivers*, Butterworth; fig. 4.15, Gregory, K and Walling, D (1973), *Drainage basin form and process*, Edward Arnold; fig. 4.18, *Hydrological data UK, 1991 yearbook*, Institute of Hydrology/British Geological Survey; table 5.1, Schumm (1977), *The fluvial system*; Wiley, table 6.1, Clowes, A and Comfort, P (1987), *Process and landform*, Oliver & Boyd; fig. 11.15, Alice Coleman in King's College, London (1985) *Utopia on Trial*, Hilary Shipman; fig. 6.4, Komar, P D (1976), *Beach processes and sedimentation*, © 1976, reprinted by permission of Prentice-Hall Inc., Englewood Cliffs, NJ; figs 6.12, 7.8, 7.16 and 7.17, Waugh, D (1990), *Geography: an integrated approach*, Nelson; fig. 7.4, O'Hare, G (1988), *Soils, vegetation, ecosystems*, Oliver & Boyd; fig. 7.5, Job, D (1989), 'Investigating soil erosion by water', *Geography Review*; fig. 7.9, Small, T (1972), 'Morphological properties of driftless area soils relative to slope aspect and position', *The Professional Geographer*; fig. 7.10, Salisbury, E (1925), 'Note on the edaphic succession of some dune soils with special reference to the time factor', *Journal of Ecology*; fig. 7.20, *The Ozone Project*, Watch/RSNC Wildlife Trusts Partnership, 1991; fig. 7.21, Dowdeswell, W (1984), *Ecology: principles and practice*, Heinemann; fig. 7.22, redrawn, with permission of Claire Dalby, after colour plates in Richardson, D H S (1992), *Pollution monitoring with lichens*, The Richmond Publishing Co. Ltd; table 8.2, Gates, E S (1972), *Meteorology*, Harrap Chambers; fig. 8.11, Fitzgerald, P (1973), *Weather in action*, Methuen & Co.; table 9.1, Thompson, D L (1963), 'New concept: subjective distance', *Journal of Retailing*; table 9.2 and figs 9.6 and 9.8, Gould, P and White, R (1974), *Mental maps*, Penguin; fig. 9.3, Lynch, K (1960), *The image of the city*, MIT Press, Mass.; fig. 9.4, Orleans, P (1967), reprinted with permission from *Engineering and the city*, National Academy of Engineering, Courtesy of the National Academy Press, Washington, DC; fig. 9.9, Fines, K D (1968), 'Landscape evaluation: a research project in East Sussex', *Regional Studies*; fig. 9.10, Essam, J (1990), 'Homes v. wildlife', *Geographical Magazine*; figs 9.13, 11.27 and 11.28, Ashworth, G J (1984), *Recreation and tourism*, Bell & Hyman; fig. 9.14, Parkes, D N and Thrift, N J (1980) *Times, spaces and places*, Wiley; fig. 10.2, The Controller, HMSO, and Charles E Goad Ltd; fig. 10.4, reproduced with the permission of Ronald Johnson, Vicar of Dorney; table 11.3, Greasley, B (1984), *Project fieldwork*, Collins Educational; tables 11.5, 11.6 and fig. 11.25, CSO *Social trends 1993*, Source: Central Statistical Office, Social Trends 23; fig. 11.4b, Hoyt, H (1939), *The structure and growth of residential neighborhoods in American cities*, Federal Housing Administration, Washington, DC; fig. 11.4c, Mann, P (1965), *An approach to urban sociology*, Routledge; fig. 11.5, Boal, F W (1969), 'Territoriality at the Shankill–Falls divide', *Irish Geography*; fig. 11.6, Office of Population Censuses and Surveys, Census 1981, HMSO; figs 11.16, 11.18, 11.19, 11.21, 11.22, 11.23 and 11.24, Tidswell, V (1976), *Patterns and process in human geography*, University Tutorial Press; fig. 11.30, Carolyn M Harrison in Slater, F (1986), *People and environments*, Collins Educational; fig. 11.31, Cox, B C and Moore, P D (1983), *Biogeography: an ecological and evolutionary approach*, Blackwell Scientific; fig. 12.12, Daniel, P and Hopkinson, H (1979), *Geography of settlement*, Oliver & Boyd; fig. 12.17, Coppock, J T (1965), 'The cartographic representation of British agricultural statistics', *Geography*; fig. 12.18, Keeble, D (1990), 'High technology industry', *Geography*; fig. 12.19, Jones, P (1991), 'The French census 1990: the southward drift continues', *Geography*; fig. 12.20, Source: *1991 Census, County Monitor, Inner London*; table 14.1, Duncan, O D and Lieberson, S (1959), 'Ethnic segregation and assimilation', *American Journal of Sociology*, © University of Chicago Press; appendices 1 and 2, Lindley, D V and Miller, J C P (1953), *Cambridge elementary statistical tables*, Cambridge University Press; appendix 3 is taken from Fisher, R A and Yates, F (1953), *Statistical tables for biological, agricultural and medical research*, published by Longman Group UK Ltd.; appendix 4, Siegel, S (1956), *Nonparametric statistics for the behavioral sciences*, McGraw-Hill, New York; appendix 5, from McCullagh, P S (1974), *Data use and interpretation*, © Oxford University Press, reprinted by permission of Oxford University Press.

We would like to thank former colleagues from whom we have borrowed ideas, particularly Alasdair Paine's on rivers, coasts and methodology and Mike Town's on Coasts and meterology.
Barnaby Lenon and Paul Cleves

Every effort has been made to trace copyright holders and to obtain their permission for the use of copyright materials. The publishers will gladly receive any information enabling them to rectify any error or omission at the first opportunity.

**Photographs**
Cover picture, NASA/USGS; fig. 7.18, V. J. Matthew/Shutterstock; fig. 10.1, Centre for Ecology and Hydrology; fig. 10.3, NRSC Ltd/Science Photo Library. All other photographs were taken by Barnaby Lenon and Paul Cleves.

# Contents

# To the student

We have tried in this text to do three things. First, to produce a full account of the wide range of types of data measurement and analysis that will be needed for A-level geographers studying syllabuses from OCR, AQA, Edexcel and WJEC from 2016 onwards. Second, to present these in a form that can be readily understood by those with little or no previous experience of such techniques. Third, to give a good helping of fieldwork report examples that might help provide some answers to the question 'What investigation can I do?'

Every investigation goes through the three stages of collecting, processing and presenting information, so we have divided the book into three in exactly the same way.

## Part A: Collecting the information

When statements are made such as 'The average precipitation in London is sixty centimetres per annum', or 'The population of Newcastle has fallen by five per cent in the last decade', it is very easy to accept these as firm facts. Many people have been employed in collecting the data upon which such conclusions are based, but the reliability of the data has to be assessed in the light of what we know about the methods by which they were collected, methods that are never perfect.

Geography depends on high-quality information from which conclusions can be drawn. Part A looks at some of the ways in which you can collect data and be sure that it is reliable.

## Part B: Processing the information

Once data has been collected it is often unmanageable in its raw form – just lists of numbers or questionnaire answers. In order to make the data intelligible we summarise it by using a map, diagram or summary statistic (such as the average of a set of numbers). But maps and statistical methods do more than just summarise data: they can reveal important patterns and relationships that are almost impossible to detect by simply gazing at the raw data. This is why they always form a part of any geographical investigation. The difficulty is in deciding which methods are suitable for the data you have collected, and this is the subject of Part B.

## Part C: Presenting the information

No matter how thorough your fieldwork or how well you have analysed your results, you will let yourself down if you do not present your project in a suitable manner. It is all too easy to forget to say what sampling technique you used and why, for example, or to begin to analyse parts of your data before you have fully stated your hypotheses. Part C should help you to avoid these pitfalls and enable you to present a worthwhile fieldwork investigation.

*Barnaby Lenon*
*Paul Cleves*

# 1  Fieldwork investigations

## 1.1 Types of fieldwork

Confusion sometimes arises when we talk about fieldwork because the word is used to cover three different types of activity.

Firstly, there is the *guided tour* type of fieldwork, where we are taken around an area and told about it. This type of field trip is valuable: indeed, there is no substitute for actually visiting the places we are studying. But the important point about the guided tour is to know as much as possible about the place before visiting it: the more you know, the more you see.

Secondly, there is true *experimental* fieldwork, which is trying to find answers to previously unresearched problems. The result of the research is not known before the work begins. This type of fieldwork is common at higher levels within universities.

Thirdly, there is *pseudo-experimental* fieldwork, where measurements are taken but the outcome is probably already known. This fieldwork is much like experiments done in school chemistry – a training in methods of data collection, but not really an experiment at all. A good

deal of geography fieldwork is of this sort: measurements are taken of all the characteristics of a central business district (CBD), for example, in order to illustrate how large or significant it is. This is not original research, by any means, and does not have to be. Some of the conclusions may tell you things about the CBD that you did not know, while others you will probably have known already but are now able to confirm through your researches. At the very least, the process of data collection, the ordering of your results, and the conclusions that you draw are worthwhile exercises in themselves.

There is, in fact, a fourth type of research: the *controlled experiment*, where research is done within an artificially controlled environment. For example, one might test rock weathering by putting samples of rock in an oven which is switched on and off to simulate the effect of day and night. Such controlled experiments can be of the true or the pseudo-experimental type.

## 1.2 How to choose a title

The most difficult but important part of a project is deciding at the outset exactly what to study. Thought and preparation at this stage will save you time and frustration later when you are trying to collect information for your project. It is important to choose a topic that interests you but, equally important, it should be one capable of producing a worthwhile result.

**Three types of individual studies tend to be particularly successful:**

**1**  Those that study a local issue or land-use conflict, such as the environmental impact of tourists or the consequences of a housing redevelopment scheme or a planned bypass. Figure 1.1 summarises the stages involved in a study where there are different groups with different attitudes and values.

**2**  Those that test a hypothesis or theory, such as von Thünen's theory that the intensity of agricultural land use falls with increasing distance from a market. Many aspects of geography have existing theories attached to them, and these can be read in standard textbooks. Or you may like to devise a hypothesis of your own.

**3**  Those that compare the geographical characteristics of two places or phenomena, such as two shopping centres or two streams. A variation on this theme is a comparison of the geographical characteristics of one

place or phenomenon at two or more stages of time, such as the population structure of a village in 1850, 1950 and 2000.

**Think through the complete range of possible topics.**
Consulting your course syllabus may help you here. Alternatively, consider the following selection.

**Physical geography topics**
*People–environment topics*
– perception of flood hazards (related, perhaps, to frequency and severity of past flood events)
– cliff/slope instability and the perception of residents to the threats posed
– fluctuations in atmospheric pollution and the effects on asthma sufferers
– patterns of atmospheric pollution within a town
– patterns of pollution along a stretch of urban river
– the effect of tourism pressure on a local beauty spot
– conflicts of interest over the deforestation/ afforestation of a local area
– effects of erosion of footpaths in an upland district (and measures for alleviating the problem)
– costs and benefits of building sea defences along a line of coast

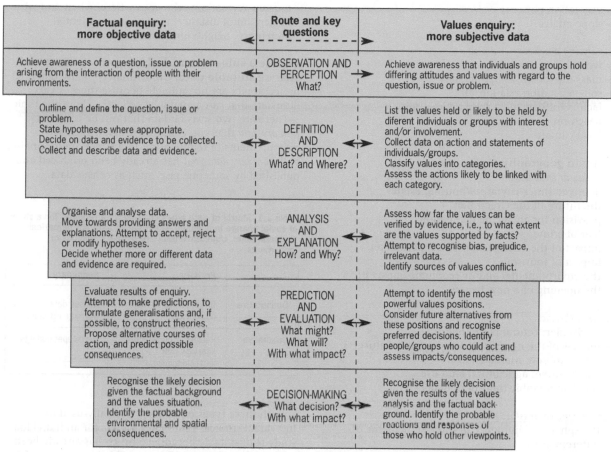

| Factual enquiry: more objective data | Route and key questions | Values enquiry: more subjective data |
|---|---|---|
| Achieve awareness of a question, issue or problem arising from the interaction of people with their environments. | OBSERVATION AND PERCEPTION What? | Achieve awareness that individuals and groups hold differing attitudes and values with regard to the question, issue or problem. |
| Outline and define the question, issue or problem. State hypotheses where appropriate. Decide on data and evidence to be collected. Collect and describe data and evidence. | DEFINITION AND DESCRIPTION What? and Where? | List the values held or likely to be held by diferent individuals or groups with interest and/or involvement. Collect data on action and statements of individuals/groups. Classify values into categories. Assess the actions likely to be linked with each category. |
| Organise and analyse data. Move towards providing answers and explanations. Attempt to accept, reject or modify hypotheses. Decide whether more or different data and evidence are required. | ANALYSIS AND EXPLANATION How? and Why? | Assess how far the values can be verified by evidence, i.e., to what extent are the values supported by facts? Attempt to recognise bias, prejudice, irrelevant data. Identify sources of values conflict. |
| Evaluate results of enquiry. Attempt to make predictions, to formulate generalisations and, if possible, to construct theories. Propose alternative courses of action, and predict possible consequences. | PREDICTION AND EVALUATION What might? What will? With what impact? | Attempt to identify the most powerful values positions. Consider future alternatives from these positions and recognise preferred decisions. Identify people/groups who could act and assess impacts/consequences. |
| Recognise the likely decision given the factual background and the values situation. Identify the probable environmental and spatial consequences. | DECISION-MAKING What decision? With what impact? | Recognise the likely decision given the results of the values analysis and the factual background. Identify the probable reactions and responses of those who hold other viewpoints. |

**Figure 1.1 The various stages involved in undertaking a geography project or individual study. The left-hand column refers to a conventional project based on objective data (such as a** study of urban climate), while the right-hand column is a scheme for a study of the conflicting views behind a land-use decision (such as a plan to build a new airport).

– the environmental impact of a new golf course
– the environmental and ecological effects of excessive water abstraction from a river

*Compare contrasting places*
– two stretches of coast
– two rivers or different parts of one river
– the local climate of two slopes (of varying aspect), in and out of a town (urban climate), in and out of a wood
– soils along a catena
– a succession of sand dunes
– two different rock types

*Change through time*
– Note that the timescales involved can vary from a few days to a period of months or even years. For the shorter time periods you can collect all the data yourself. For the medium timescale you may need to use secondary data such as river flow data, and for the longer term you will need old maps, photographs

and other documents. Of course, it is possible to compare your primary data with secondary data collected years before, but beware of variations caused by differences in method. Note, too, the *ergodic hypothesis* (see page 10), which suggests that space may be substituted for time. For example, the process of coastal erosion may be more advanced in one area than in an adjoining one
– variations in the intensity of the microclimate of a wood under different synoptic conditions
– variations in beach profiles in summer and winter
– the effect of urban growth or forest clearance on the discharge or sediment load of a river
– patterns of coastal advance or retreat
– migration of river meanders

*Analysing distributions*
– glacial landforms
– drainage patterns
– sediment characteristics
– meanders

– vegetation patterns
– slope angles

*Analysing processes*
– weathering, erosion
– mass movement, slope stability
– soil–vegetation relationships
– factors influencing a river hydrograph
– beach processes
– local climate

## Human geography topics
*Looking at local issues*
– deprived inner-city areas/housing estates
– the future of the Green Belt
– provision for an ageing population
– fear of crime
– impact of the construction of a supermarket
– impact of tourism
– the environmental and/or economic impact of
  the opening of a new bypass

*Testing theories*
– Christaller: central place theory
– models of the internal structure of cities: Burgess,
  Hoyt, Robson, Mann, Alonso
– von Thünen: agricultural land use
– the demographic transition model

*Comparing contrasting places*
– the spheres of influence of three settlements of
  different sizes
– two or more shopping centres
– two or more urban neighbourhoods

*Change through time*
– population characteristics of a parish over a
  hundred years
– impact of social change on a rural village
– land use in an urban neighbourhood
– gentrification of a small inner-city area
– shop types in a High Street

*Analysing distributions*
– ethnic groups
– leisure centres
– schools
– hospitals, clinics, GPs
– old people's homes
– types of housing tenure
– crime
– noise
– disease and illness

*Studying the behaviour of people*
– shopping behaviour
– household movement

– tourism/recreation: who goes where, when and why
– perception of distance, landscapes, residential
  preference, neighbourhood boundaries

**Choose a subject that enables you to collect and analyse suitable data.** Unfortunately, many potential project topics are not suitable because there is very little data that can actually be collected about the topic.

There are two sorts of data that can be collected: *primary data* that you collect first hand, such as measurements of temperature or a street questionnaire, and *secondary data* that has already been produced and published by someone else, such as census data.

Table 1.1  Matrix of data types showing data required for a study of social change in a village. Quantitative data are numerical measurements. Qualitative data are people's opinions or judgements.

| | Primary | Secondary |
|---|---|---|
| **Quantitative** | Street survey to map house type | Census data Marriage registers |
| **Qualitative** | Residents' questionnaire Interview with local clergyman | Newspaper cuttings |

Both these types of data are normally used in individual studies. Sometimes a topic for an individual study is unsuitable because so much has already been published on that specific topic. This is one reason for avoiding a straightforward descriptive study of, say, London Docklands – so much will already have been published by the former Development Corporation.

All examination boards place a good deal of emphasis on the data-collection side of the individual study. For this reason it is often wise to come up with a project topic that enables you to use a wide range of data sources, not just one. Projects based solely on questionnaires, for example, can be successful but are harder to do well than those that use a wider range of data sources.

One category of project that deserves to be treated with caution is the *impact study*, such as the impact on an area of building a road. This type of project can be very successful but suffers from the problem that, in the timescale under which most A-level projects work, it is usually impossible to collect data both before and after the road was opened. You cannot really study the impact of a scheme like a new road if you cannot obtain such data.

**Choose a fairly narrow subject.** It is often better to study one stream in detail than five streams sketchily, one shopping street rather than several.

The size of the area studied will depend on the topic: studying settlement distribution and settlement hierarchies means that you need to look at an area large enough to contain a good number of settlements of a variety of sizes. On the other hand, a study of social change in a city might well be conducted at a scale no bigger than half a dozen streets, especially if detailed census data is available. A microclimate project looking at the variations in climate around just one building is perfectly acceptable and often more fruitful than a more generalised study of a whole town.

Sample size (page 16) is a consideration here. It is important that the area studied should be representative enough to enable you to draw conclusions that can, in all likelihood, be applied to other similar areas. This is why a study of one factory or one farm is unlikely to be acceptable: it is unsound to draw general conclusions from such a small sample size.

**Choose a topic that is geographical:** in other words, that deals with distributions, the characteristics of places, or the relationship between people and the environment. Descriptions of factory processes or the workings of a farm are the sorts or themes that are not acceptable.

**Be led by certain additional considerations:**
1 Which topics have particularly interested you in your course?
2 Would any of your other subjects help? For instance, if you are studying chemistry you might be interested in examining the chemical composition of different horizons in a soil profile.
3 Have you any special interests or hobbies that might be relevant? If you are a keen angler, for example, you may wish to investigate stream pollution.
4 Do you have any special contacts or sources of information? You may have access to the records of a family business, for example, which might be included in a study of the changing location of markets for a particular commodity.
5 Is there anything interesting happening in your local area? Topics such as controversial new road schemes, the siting of a proposed waste dump, or a new regional shopping centre might be investigated.

If you still cannot think of a title for a suitable study, a good way of producing ideas is to have a close look at an OS map of your home area.

**Do not be too attracted by the idea of doing a project on a holiday location that cannot be revisited.** Especially if this is abroad, it can be impossible to know what the data sources are like before you go, and if you make a mistake with your data collection it is impossible to correct it.

**It is often a good idea for your project title to be defined in terms of a hypothesis or question.**
A bad project title would be *A study of Brighton* – it is too vague. Less bad is *A study of old people in Brighton*. Better still is *What factors determine the distribution of old people in Brighton?* or a hypothesis: *The distribution of facilities for old people in Brighton does not match demand.*

Here are some examples of potentially successful project topics:

- What factors influence the pattern and rates of gravestone weathering?
- How does longshore drift vary down a stretch of coast over two months?
- The morphology of glacial deposits in a Lake District valley.
- What factors control stream velocity?
- Morphological contrasts between a suspended-load and a bedload-dominated stream.
- What impact does rock type have on river morphology, flow and load characteristics?
- What is the relationship between rainfall and stream discharge in a small drainage basin?
- How and why do soil characteristics vary down a slope?
- Vegetation patterns across a salt marsh in relation to tidal and other influences.
- The effects of exposure and shelter on windspeeds at a coastal location.
- The microclimate of a small wood.
- Pollution patterns in an urban watercourse.
- Cliff profiles in relation to materials and processes.
- Infiltration rates on different soils or slopes at different seasons.
- What is the impact of tourism and recreation on Ashdown Forest?
- The social and economic consequences of flooding and flood control in the Maidenhead area.
- Impact of coastal erosion on the residents of the Black Ven area, Dorset.
- The impact of afforestation on the Kielder area.
- How has the CBD of Sunderland changed since 1980?
- Testing central place theory in south Norfolk.
- A comparison between the shopping behaviour of two neighbourhoods.
- What factors have influenced the morphology of Crewe?
- The factors behind recent industrial location decisions in Swindon.
- Mental maps of an urban area and reasons for the variations seen.
- The relationship between the design of council estates and patterns of litter, vandalism, graffiti and other indices of decay.
- The distribution of crime in east Bradford in relation to housing and other conditions.

- The distribution of the Jewish population of Barnet.
- Rates of population turnover in parts of Slough.
- Variations in the distance between the premarital addresses of marriage partners over time.
- Testing the demographic transition model on a parish village over 100 years.
- The influence of a city on farming in the rural–urban fringe.
- The environmental effects of water abstraction from the River Pang in Berkshire.
- Kerbside patterns of carbon monoxide pollution in Kettering.
- The effect of pollution on asthmatics in Rochdale related to prevailing climatic conditions.
- Coastal change in Purbeck in the twentieth century: the evidence of maps, photographs and postcards.

- Patterns of pollution along the River Irwell as measured by biological indicators.
- Seasonal variations of microclimate intensity in the Forest of Bowland.
- The perception of the inhabitants of Selby to the risk of flooding. (This compares not only the distance of residents' homes from the River Ouse but also the length of time they had lived in Selby, housing tenure, building type and other variables.)
- The Malham honeypot: measuring the impact of visitors on an area of outstanding natural beauty.
- Sediment changes in the River Taff during construction of a superstore.
- An investigation into patterns of carbon monoxide pollution along the M8 in Glasgow.

## 1.3 Investigation method

There is a five-fold classification of different methods that you can use when doing a project:

### Direct observations of processes

These are projects where you make a direct observation of the processes at work, such as a questionnaire designed to find out why people go shopping where they do.

### Mapping and correlation

Many projects measure or map two or more variables and attempt explanations on the basis of correlations observed. For example, you might measure beach angle and the particle sizes of sediments on the beach and note the correlation between the two.

The advantages of this method are that it is both simple and a useful first stage in research. The difficulty is that a correlation between two variables leaves unanswered a number of questions about processes. In the example given above, we cannot be sure whether particle size determines beach angle, or beach angle determines particle size, or whether they are in fact both influenced, independently, by a third factor such as wave energy. In nature causality is often complex, with partial correlations between a host of variables. We must therefore be very wary of concluding too much from simple correlations.

### Reconstruction methods

If we are studying change over time it is usually necessary to reconstruct the past, perhaps by using Kelly's Directories to show patterns of shops in a High Street fifty years ago. Examples from physical geography include till fabric analysis (for glacial deposits) or the reconstruction of river meander shifts, which you can do by using a sequence of old maps.

### Simulations

In the natural world change can be too slow to observe. An effective way of studying change, therefore, can be to make a scaled-down version of reality such as a flume for studying river evolution, or to use a fridge and oven to simulate the influence of diurnal heating and cooling on rocks. You obtain results quickly, and such a method enables you to experiment with variables – for example, the impact of different slope angles on river form.

### Location-for-time substitution

This is something called the *ergodic method* and is another way of studying change over time. The method involves finding places that are at different stages of development and, by comparing them, learning about the process of change they may go through. For example, to study gentrification one might well compare three neighbourhoods that are similar, except that one started to experience the arrival of the middle classes twenty years ago, another only five years ago, another hardly at all. Another example is the coast of Thanet, where sea walls have been built along the base of the chalk cliffs at various times over the past 40 years. Once cliffs are protected from sea erosion their angle begins to decline, and a model of evolution may be hypothesised. Beware of assuming that because a particular sequence occurs in one place it will necessarily occur elsewhere.

## 1.4 How to carry out a fieldwork study

1 Start by being clear about what you are trying to do. This normally means defining your aims in terms of one or more hypotheses or questions.

2 Your project will be divided into three stages:

a *Collecting the data*. At this stage you should be concerned about three things:

i The data sources, which might be primary (such as questionnaires, river measurements, temperature readings) or secondary (census data, local authority town plans, newspaper reports).

ii The sample size: how many questionnaires should you collect to gain a representative set of answers, how many points on the river should you measure?

iii The time when you will collect the data. If you are doing a survey of shoppers, the time of day and day of the week affect the sort of people who are out in the High Street, so the time chosen for your survey matters. If you are doing a project on local climate, the time of year and the prevailing weather conditions will affect your results.

b *Summarising and analysing the data* using maps, diagrams, statistics and methods of spatial analysis.

c *Describing these results in words.*

3 Ensure that you are organised and have thought through likely problems before you begin to collect data. This is particularly important with studies that require equipment as well as the preparation of tables on which to record results. Questionnaires should be carefully planned and tested.

4 Be flexible because problems always arise: rivers dry up, people refuse to answer questionnaires, and other sources of data may simply fail to come to anything.

5 Be sure to study processes as well as patterns. If, for example, you study land uses in a town or the distribution of different types of shops, it is important to try to explain why these patterns have evolved. This might mean consulting the local history section of a library and using questionnaires and interviews, as well as some intelligent guesswork.

6 Given the wide range of techniques available to you, it is important to think carefully about which one to use. Try to weigh up the pros and cons of each method as it affects you, and be prepared to discuss this in the 'techniques' section of your write-up.

---

## 1.5 Fieldwork safety

Safety in the field applies equally whether you are doing a physical or human project. Obviously, the dangers against which you must guard vary from town to country. As a first step, your teachers or parents must approve of the fieldwork you intend to undertake and know where you will be working as well as the dates and times.

*Know the risks*

The risks inherent in fieldwork locations are summarised in Table 1.2.

**Table 1.2  Potential fieldwork risks** *(adapted from University of Cambridge guidelines)*

| Physical hazards | Man-made hazards |
|---|---|
| Extreme weather | Road and rail traffic |
| Remoteness | Machinery, vehicles |
| Mountains and cliffs | Power lines, pipelines |
| Glaciers, crevasses, ice falls, etc. | Electrical equipment |
| Caves, mines, quarries | Insecure buildings |
| Forests (including fires) | Slurry and silage pits |
| Freshwater | Attack on person or property |
| Sea and seashore (tides, currents, etc.) | Military activity |
| Marshes, quicksand | |
| Aggressive or poisonous animals | |
| Agrochemicals, pesticides | |

*Be prepared*

Preparation for fieldwork means understanding the potential risks that may arise and being adequately equipped and able to cope with them. As a general rule it is advisable to work with another person. Not only does this speed up the process of data collection (recording while you observe, for instance) but it means that help is at hand should you find yourself in difficulty – cut off by the tide or lost in a remote and unfriendly part of town. But fieldwork could be no more than interviewing fellow pupils in school or measuring temperatures in your garden, in which case the need for safety hardly arises.

Some forms of preparation that may be relevant to you are listed in Table 1.3. You need to get to the site and have permission to work there, and you need to have the right equipment. Permission to work in an area is obviously important but sometimes overlooked, as it is sometimes not easy to know who the owner of the site is. Use your discretion: if you are measuring stream discharge, for instance, you are unlikely to disturb the farmer; but if you need to dig soil pits or leave max–min thermometers in an orchard then you should obtain permission. Nowadays, shops and shopping centres are reluctant to allow questionnaire surveys on their property, and many shopping malls forbid them outright.

**Table 1.3 Preparation for fieldwork** *(adapted from University of Cambridge guidelines)*

| Access | Equipment |
| --- | --- |
| Travel arrangements | Safety clothing, e.g. hard hat, |
| Permission for access to sites | waterproof clothes |
| Availability of assistance | First-aid kit |
| Accommodation | Survival kit, e.g. whistle, bivvy bag |
| | Emergency food and drink |
| | Navigation aids: maps, compass |
| | Mobile phone |

*Precautions*

Some advice and common precautions are spelt out below.

- Ensure that your parents know where you are working and when you are coming home.
- By law, safety helmets must be worn when visiting quarries, mines, building sites, cliffs and screes, or wherever there is a risk of falling objects.
- Wear reflective clothing when on the streets after dark.
- Carry warm and waterproof clothing if working in upland areas.
- Keep an eye on the weather and tides.
- Beware of slippery river banks, fast-flowing streams and deep pools.
- Be polite when interviewing the public, and do not cause offence in any way.
- Do not damage property, close all gates that you pass through, do not trample crops or worry livestock.
- Do not collect specimens from nature reserves without a permit.
- Do not leave litter.
- Respect wild life. Beware of adders in heathland.
- Always obtain permisson before entering private property.
- Do not work alone, particularly in fluvial, estuarine, coastal and upland areas, and also in urban areas and if interviewing people in their homes.
- Do take adequate food and drink.
- Use a good map and a hand-held GPS to minimise the chances of getting lost.

# 2 Sampling

## 2.1 Why do we sample?

If we wanted to examine the shopping habits of people in a middle-sized town it would not normally be possible to interview all the residents: we would have to sample a cross-section. However, it is not always necessary to sample. If the total number of items under consideration, known as the *population*, is small, we might be able to measure them all. In fact, where the total population is less than 100 it would normally be wise to measure all of them: a sample of such a small number is likely to be unreliable.

But often sampling *is* desirable, for several reasons:

1 It is *quicker* than measuring every item.
2 It is therefore much *cheaper*.
3 Often it is *impossible to measure everything because the population size is so great*. If we wished to find out the average size of pebble on a beach, for example, we would obviously not be able to measure each pebble.
4 It is *unnecessary* to measure the whole population because a carefully chosen sample can give you a result (such as average pebble size) that is close to the figure you would obtain if you measured every pebble. There are techniques available that tell you how close your sampled result is likely to be to the result you would obtain if you measured every one.
5 It is sometimes *impossible to gain access* to the complete population. For example, if we did a questionnaire survey of people's shopping preferences in a town, many would refuse to answer the questionnaire. We can only measure a proportion of the population, making the sample biased against those types of people who are unwilling to co-operate (such as adults accompanied by small children).

6 We may wish to *take a snapshot* of the population at one moment in time. In this case we would be forced to sample because it would take too long or need too many people to measure the whole population. For example, meteorological readings are taken simultaneously at only a limited number of sample sites in the UK and sent to the Meteorological Office, as it would not be possible to record weather at every spot in Britain at the same time. To take another example, if we were recording the flow of traffic down the streets of a town it would be no good recording one street at 10 a.m. and another at noon because the two streets cannot then be fairly compared: traffic flows at noon may be quite different from those at 10 a.m., and we would not therefore be able to tell whether differences in flows between the two streets were due to the different times at which the flows were recorded or due to any other contrasts in the character of the streets. For a fair comparison the traffic flows should be measured simultaneously in the two streets, and normally that will force us to sample.
7 We *may not know the population size or location*, in which case we will be forced to sample. For example, if we are investigating the distribution of the Jewish community in a small city we might well be forced to sample because there is no data source available that enables us to find the addresses of Jewish people in the UK. A project on the distribution of foxes in part of a city is bound to sample because we do not know how many foxes live there in the first place.

## 2.2 Avoiding bias

When deciding how to take a sample and how many samples to take, the main consideration is obtaining a representative result – avoiding bias. Bias can arise for many reasons:

1 The data from which the sample is taken may themselves be biased. Selecting the addresses of, say, 100 people from a telephone directory is biased against lower-income people who do not have a private telephone. The data from which the sample is drawn are called the *sampling frame*.

2 Insufficient care in the choice of sample may result in an unrepresentative result. For example, to obtain a picture of shopping behaviour, it is necessary for us to issue a questionnaire to a representative number of old as well as young people, men as well as women.
3 The time the sample was taken may produce bias. Conducting a street questionnaire during the morning of a weekday is biased against working people and biased towards those who do not work, including retired people.

## 2.3 Sampling methods

Before selecting a sample it is necessary to decide on the sampling method that will give the most accurate result with least effort, and the sample size.

When selecting samples from an area on the ground or from a map there is one preliminary decision to be made (Fig. 2.1): are you looking for points, lines or quadrats?

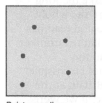 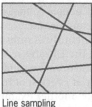 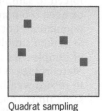

Point sampling    Line sampling    Quadrat sampling

**Figure 2.1  The three spatial sampling methods**

*Point sampling* involves choosing individual points and sampling at those points, such as specific houses down a street or crop types on a land-use map.

*Line sampling* involves taking measurements along a line. For example, to sample vegetation across a series of sand dunes we might lay a tape across the dunes and note the dominant vegetation type along each part of the line.

*Quadrat (or area) sampling* involves marking a square on the ground or on a map and noting the occurrence of the feature you are interested in within the square. For example, if you were interested in the geological make-up of pebbles on a beach, as you might be if studying longshore drift, lay a quadrat on the beach and count the number of pebbles of each known rock type within the square. This is more likely to obtain an unbiased result than attempting to pick individual pebbles by point sampling or line sampling.

How do we decide where to put our points, lines or quadrats? How, if we are doing a street survey, do we decide which people to stop and question? How, if we are selecting names from a directory, do we choose those names? It is no good just closing our eyes and stabbing at the map or directory with a pencil because such a method may well result in bias. There are three commonly used methods. To illustrate these we shall use the example of an area composed of two rock types (A and B) on which we wish to sample land use (Fig. 2.2).

### Random sampling

Random sampling has a specific meaning: it means sampling using random numbers. The method in our example is as follows:

1 Decide how many sample points you want.
2 Obtain *random numbers*, either from a published sheet (page 160) or from a calculator that generates random numbers. Random numbers are what they sound like: numbers that are chosen at random.

3 Overlay the map with grid lines and number them.
4 Use the random numbers to read off grid references. Plot these on the map (Fig. 2.3). The first numbers in the top left-hand corner of the table are 20 17. We make 20 the easting grid reference and 17 the northing grid reference. This becomes a point on Figure 2.3. This procedure has been followed for twenty points.
5 Find out the land use at each of the points chosen by visiting the places on the ground using a hand-held GPS.

If we are doing line sampling rather than point sampling, we use the random number tables to give us two points on the axes of the grid; these are joined to produce a line.

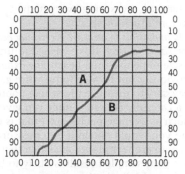

**Figure 2.2  An area of land with two rock types, A and B. The map is covered with a grid.**

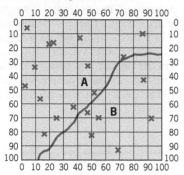

**Figure 2.3  Random sampling: 20 points have been plotted on the map using random numbers**

Random numbers can be used for most types of sampling. If we were doing a household questionnaire, the numbers 20 17 would tell us to visit the second house along, then one after that, then seven after that. If we were measuring pebble size on a beach we might lay out a tape measure and take pebbles at the two-metre mark, then one metre after that and seven after that.

The merit of using random numbers is that they provide a means by which we can select samples in the knowledge that no human bias is involved in the selection process.

The disadvantage of the simple type of random sampling described is that if the sample size is small we might, by bad luck, obtain an unrepresentative result. If we plotted only ten points on Figure 2.3, for example, it is perfectly possible that they might all fall into the area occupied by rock type A. This is why we usually prefer to use stratified sampling.

### Stratified sampling

With stratified sampling we start by asking the question: are there subsets of the pattern being measured that must be included within our sample? In a project about the decisions made by farmers, for example, we might well believe that young farmers behave differently from old and farmers with large estates behave differently from small farmers. We would therefore make sure that our sample included a representative proportion of young and old, large and small farms. These subsets are called 'strata'.

In our study of land use (Fig. 2.2) we can be fairly sure that rock type is an influence, and we must ensure therefore that a fair proportion of our sample comes from each of the two rock types. If rock type A takes up 60 per cent of the area it should receive 60 per cent of the sample points, i.e. twelve out of the total (Fig. 2.4). The other eight points will be allocated to rock type B. Plot the points as for random sampling (above), but do not allow more than the allotted number of points to fall on each rock type. Ignore any excess points that are chosen by the random numbers.

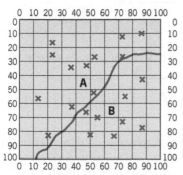

Figure 2.4 Stratified sampling: 20 points have been plotted, 40 per cent (8 points) in area B, which is 40 per cent of the area

Stratified sampling has the great advantage that it helps to reduce any bias that might possibly arise if samples were chosen completely at random. The only problem is identifying what the strata should be. If we were doing a study of shopping behaviour we might believe that all sorts of characteristics are relevant: age, sex, ethnicity, income, occupation. The difficulty is knowing for certain that these are the relevant strata to use, and how many strata to create within each (how many age divisions, for example).

### Systematic sampling

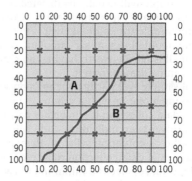

Figure 2.5 Systematic sampling: 20 points have been plotted on the map in a regular pattern

Here we choose our sample according to some agreed interval – every fifth house, or every tenth pebble lying along the measuring tape. Figure 2.5 shows how we place our 20 points. We might equally place lines or quadrats at regular intervals over the area.

The advantages of this method are great. It ensures complete coverage of the map and is simple to do – simpler than using random number tables. The danger is that the systematic sample might inadvertently pick up some underlying regularity. For example, a systematic sample down a stretch of beach, where all measurements were taken close to one of a series of regularly spaced breakwaters, would result in bias.

It is possible to combine stratified and systematic sampling methods. First decide on the strata (for example, high-, middle- and low-income areas) and then choose systematically within them.

Figure 2.6 Points, lines and quadrats plotted on a map using random, stratified and systematic sampling procedures. The area is divided into two rock types and rock type influences land use, the subject of the study.

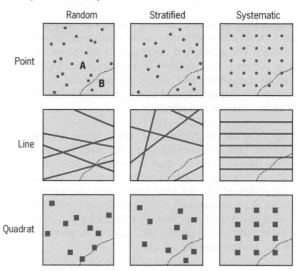

## 2.4 Sample size

The size of a sample will usually be dictated by the time and resources available. It is important to realise that the larger the sample the more likely it is to give a true picture of the population you are sampling. A sample that is too small is hardly worth having: no realistic conclusions can be drawn from it.

The size of sample will partly depend upon whether you are working alone or with a group. The extent to which group data collection is allowed as part of the contribution towards an individual study will vary according to the examination syllabus. Remember that some techniques of data collection need more than one person: a slope survey needs a minimum of two people (both hold the range poles, and one reads the clinometer), and a traffic count that needs to be taken at several parts of a town at the same time may need many assistants.

Another factor that influences the sample size is your capacity to handle the data collected. Statistical packages for computers, which enable you to handle a large amount of data quickly, should allow you to collect much greater sample sizes.

Where your sample survey involves making numerical measurements, a technique exists for calculating the minimum sample size needed. Let us imagine we wish to estimate the average length of all the pebbles on a beach. We cannot possibly measure all the pebbles, but what number should we measure to be sure that our result is close to what we would obtain if we did measure all the pebbles?

*Stage 1* Collect a pilot sample of 30 pebbles. Measure their lengths.

*Stage 2* Calculate the standard deviation (*s*) of the pilot sample results. The procedure for doing this is described on pages 142–3.

*Stage 3* Decide on the tolerable margin of error (*d*) – how close you wish to be to the result you would obtain if you measured every pebble on the beach (e.g. 0.5 cm). The smaller the margin of error you choose the greater the sample size you need to pick up, so it is important that you do not choose margins of error that are too small.

*Stage 4* Decide on the level of certainty you wish to achieve. This is expressed as a percentage figure. You might decide that you wish to be 90 per cent certain that the average length of pebble measured in your sample lies within 0.5 cm of the result you would obtain if you measured every pebble on the beach. Divide the percentage certainty figure by 200. Look up this value in Appendix A2, column A, and read off the corresponding *z*-score (*z*).

*Stage 5* Calculate the formula:

$$n = \left(\frac{zs}{d}\right)^2$$

where  *n* = required sample size
$z$ = z-score (stage 4)
$s$ = standard deviation of pilot sample (stage 2 above)
$d$ = tolerable margin of error (stage 3 above)

This tells you the required sample size. The pilot sample counts towards this total.

**Example**
To estimate the average length of pebbles at one end of a beach.

*Stage 1*
Pilot sample of 30: lengths of pebbles in centimetres

| $x$ | $x^2$ | $x$ | $x^2$ | $x$ | $x^2$ |
|---|---|---|---|---|---|
| 2.5 | 6.2 | 4.7 | 22.0 | 3.6 | 13.0 |
| 3.1 | 9.6 | 6.3 | 39.7 | 3.7 | 13.7 |
| 4.6 | 21.2 | 1.2 | 1.4 | 2.9 | 8.4 |
| 2.2 | 4.8 | 1.4 | 1.9 | 2.8 | 7.8 |
| 1.9 | 3.6 | 2.1 | 4.4 | 2.6 | 6.8 |
| 4.8 | 23.0 | 2.3 | 5.3 | 2.7 | 7.3 |
| 5.3 | 28.0 | 6.2 | 38.4 | 1.4 | 2.0 |
| 6.1 | 37.2 | 6.1 | 37.2 | 1.9 | 3.6 |
| 3.2 | 10.2 | 4.2 | 17.6 | 2.3 | 5.3 |
| 4.2 | 17.6 | 3.7 | 13.7 | 2.7 | 7.3 |
| | | | | $\Sigma x$ 102.5 | $\Sigma x^2$ 418 |

$$\bar{x} = \frac{102.5}{30} = 3.4$$

$$\bar{x}^2 = (3.4)^2 = 11.6$$

*Stage 2* Standard deviation of pilot sample (*s*)

$$s = \sqrt{\left(\frac{\Sigma x^2}{n} - \bar{x}^2\right)} = \sqrt{\left(\frac{418}{30} - 11.6\right)} = 1.52$$

*Stage 3* Decide on tolerable margin of error (*d*)
= 0.5 cm.

*Stage 4* Decide on required level of certainty = 98%
98 ÷ 200 = 0.49
= z score of 2.3

*Stage 5* Calculate formula:

$$n = \left(\frac{zs}{d}\right)^2 = \left(\frac{2.3 \times 1.52}{0.5}\right)^2 = 48.9 \text{ pebbles}$$

We have already measured 30 pebbles and need only, therefore, measure 19 more.

# 3 Geology, landforms and slopes

## 3.1 Geology

Few geographical studies have a purely geological theme, but in a number of physical and applied investigations you may need to take geology into account. At the simplest level it may be enough just to identify the rock type. Students doing hydrological or soils projects will need to know the type of rock that underlies the study area and its properties in terms of permeability, porosity and chemical composition. More specialised geological investigations will perhaps require a greater knowledge of local geology, which you can gain by using the techniques described below.

Identification of geology may be possible by observation where the rock is exposed: a cliff face, a road cutting, or where thin soils (as in chalk areas) permit. Where this is not possible, and in areas of complex geology, you may need maps of solid and drift geology. Geological maps are obtainable from good map suppliers and from the British Geological Survey website. Maps of Great Britain are available at the 1:625 000 scale (for a general indication), and at the 1:250 000, 1:50 000 and 1:25 000 scales (although national coverage at the 1:25 000 scale is patchy).

Geology includes both rock lithology (the type of rock: chalk, clay, etc.) and rock structure, the way the rock lies (folded, tilted or faulted). Rock structure is an important influence on landforms and landscape, and you sometimes need to know the local geological structure in order to interpret the landscape. You can do this either by reference to geological maps (see above), which normally include cross-sections through the map area, or by reference to local geological guides. The two most useful series of guides include that produced by the British Geological Survey, which gives national coverage, and the Classic Landforms Guides published by the Geographical Association, which are available for a select number of sites only.

### Geomorphological sites

You may be lucky enough to live within a reasonable distance of one of the UK's major landmarks such as Malham Cove or Stair Hole. Sadly, such sites are extremely popular with school and college parties and, although these are exciting features to study, you can do just as good a project on a lesser-known feature.

If you or your teachers do not know of suitable sites for fieldwork consult the British Geological Survey, www.bgc.ac.uk. Many are sites of Special Scientific Interest or Local Geological Sites, described on the Natural England website.

Many such sites are in National Parks, on National Trust land or in local conservation areas, so you must first get permission from park wardens or landowners. Treat the countryside with respect and care, and ensure that you cover over any traces of excavations you have to make.

*Discovering Landscape in England and Wales* (Goudie and Gardner, 1985), is another valuable source of ideas.

## 3.2 Weathering

The two main types of weathering are physical and chemical weathering. Physical weathering is caused by stresses created in rocks by heating and cooling, the freezing and thawing of water, wetting and drying, and the growth of salt crystals. Chemical weathering occurs when the minerals that make up the rock react to changes in the chemistry of their environment, causing the rock to dissolve or decay.

Distinctive landforms that we associate with different rock types develop partly as a result of the weathering process to which they have been subjected. Rates of weathering are controlled by climatic and microclimatic conditions. These rates also depend on the susceptibility (in terms of both chemical composition and physical properties) of different rocks to weathering processes. Because weathering takes place at such slow rates, precise measurements cannot be undertaken without expensive equipment and more time than an A-level student has available. But you could make subjective assessments of the degree to which rock has been weathered out in the field, and then make more readily quantifiable measurements of rock decay in the laboratory.

*Field observations*

### Gravestone weathering: Rahn's Index
A visual technique for determining the extent of weathering can be used on gravestones. Since gravestones are dated, you can work out the approximate time they took to arrive at a particular state of decay. Rahn's Index allows you to apply a degree of quantification to the assessment of gravestone weathering. Categories of weathering are listed in Table 3.1, and this allows you to rank stones of differing lithologies and exposure times on a scale.

**Table 3.1  Rahn's index of weathering**

| Class | Description |
|---|---|
| 1 | Unweathered |
| 2 | Slightly weathered – faint rounding of corners of letters |
| 3 | Moderately weathered – rough surface, letters legible |
| 4 | Badly weathered – letters difficult to read |
| 5 | Very badly weathered – letters almost indistinguishable |
| 6 | Extremely weathered – no letters left, scaling |

Sandstone, however, tends to flake, and in such cases Rahn's Index may be inappropriate. The use of a 50 × 50 cm quadrat allows you to estimate the area affected by flaking.

### Ollier's friability test
The friability of a rock is its crumbliness, and reflects the rock's natural hardness and the amount of weathering it has suffered. Ollier's test is a five-point scale (Table 3.2) that allows ranking of rock on the basis of its freshness and resilience to physical attack.

**Table 3.2  Ollier's scale of rock friability**

| | |
|---|---|
| 1 | Fresh; a hammer tends to bounce off the rock. |
| 2 | Easily broken with a hammer. |
| 3 | The rock can be broken by a kick (with boots on), but not by hand. |
| 4 | The rock can be broken apart in the hands, but does not disintegrate in water. |
| 5 | Soft rock that disintegrates when immersed in water. |

### Weathering of buildings
Many buildings, particularly old ones, are made of stone; the stone is usually local, although larger and more important buildings are sometimes made from stone quarried far away. Rates of weathering are often many orders of magnitude greater in cities than in rural areas because of atmospheric pollution (see Section 7.5).

Carbonate rocks (limestones) are much affected by the most serious of these atmospheric pollutants, sulphur dioxide, which is a product of burning fossil fuels. Table 3.3 summarises the different weathering processes responsible for attacking limestones.

**Table 3.3  Processes of weathering on limestone buildings**

| Name | Mechanism | Location |
|---|---|---|
| Carbonation/ solution | Water acidified by $CO_2$ dissolves $CaCO_3$. | Any water-washed sites. |
| Salt weathering | Salt crystallisation and hydration within stone exerts pressure on stone and produces breakdown. | Near salt sources, e.g. ground where road salt (NaCl) involved; below damp courses. |
| Freeze–thaw weathering | Water expands on freezing, producing stresses and breakdown of stone. | Any area where water accumulates and freezes, e.g. north-facing walls. |
| Biological weathering | Chemical and physical weathering by bacteria, algae, lichens, mosses and higher plants (e.g. ivy). Many organisms produce acids that dissolve $CaCO_3$. Also, bird droppings contain ammonia, which can attack limestone. | Bird droppings and lichens are common on sloping and horizontal surfaces. Green algal films form on north-facing walls. Bacteria are widespread. |
| Sulphation and washing | Sulphuric acid reacts with $CaCO_3$ to produce gypsum ($CaSO_4$), which is removed in solution by runoff water. | Well-washed sites where sulphur dioxide pollution is found (e.g. most city centres). |
| Sulphation without washing | Sulphuric acid reacts with $CaCO_3$ producing gypsum, which builds up with soot to form a blackened gypsum crust. | Sheltered sites (e.g. under overhanging stone) in sulphur dioxide polluted areas. |

Methods for determining the rate of surface lowering (such as micro-erosion meters) are not usually available in schools. If you want to investigate the comparative susceptibility of different building stones, you will need to use one of the laboratory methods described below. The drawback with these simulated processes is that they are not the same as the processes occurring in nature. As an alternative, you can examine the products of weathering, which give valuable clues as to the nature and rate of attack. Table 3.4 contains a list of signs you should look for. Where lead lettering or plugs were originally incorporated flush with the stone surface, they often now stand clear, revealing the amount of lowering that has taken place. Some old buildings show the date of construction – this will help you to calculate a rate of weathering. Because limestone contains fossilised shells in a calcium carbonate matrix, the stone may weather at varying rates. Shells tend to be more resistant to weathering than their matrix.

**Table 3.4 Visual symptoms of weathering processes on limestone**

**a Features produced by sulphation under sheltered conditions**
  1 A blackened crust of soot, gypsum and bits of the underlying stone.

  2 Blisters developed within gypsum crust as the crust grows in all directions, causing stresses.

  3 Large exfoliated patches developed where blistering continues – a pale, often highly eroded surface.

**b Features produced by solution/sulphation and washing**
  1 Rough surfaces develop where fossils and hard veins in the stone (which are more resistant to acid attack) stand proud of the rest of the stone surface.

  2 On vertical surfaces bedding planes (which are often less resistant to acid attack) are preferentially attacked, producing a stripey surface.

  3 On horizontal and sloping surfaces rain produces pitting.

  4 On sloping and vertical surfaces rain produces rills.

**c Features produced by biological weathering**
  NB: These features can usually only be observed when the organisms that formed them have died or have been removed.

  1 On horizontal and sloping surfaces lichens can produce small pits and eroded patches.

  2 On vertical surfaces creepers produce various indentations.

**d Features produced by salt weathering**
  White efflorescences (powdery crystals on the stone surface) of salt show that salt is present. These are often associated with flaking of the surrounding stone.

**e Features produced by freeze–thaw**
  Cracked blocks and the production of large flakes where standing water is common.

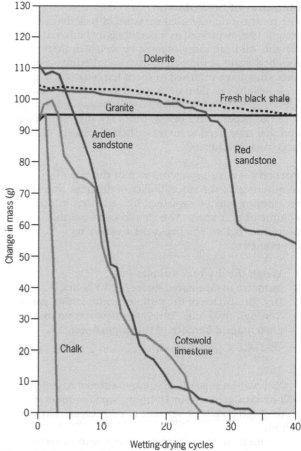

**Figure 3.1 The effect of salt weathering on different rocks**

*Laboratory methods*
Measurement of rock breakdown in a laboratory involves the simulation of weathering processes. By accelerating heating–cooling and freeze–thaw cycles, it is possible to condense months' or even years' worth of weathering into a matter of days or weeks. The exact method you use will depend upon the resources available, but accuracy and consistency of method are important, as is a careful description of the technique in the project write-up.

You have to decide on the degree of reality you wish your experiment to reflect. Some weathering processes operate in arid or semi-arid environments only. Insolation weathering in a desert rarely produces a diurnal temperature range in excess of 30 °C, but in order to produce more rapid results you may need to subject your samples to greater extremes of heat and cold. Your experimental cycles will inevitably be shorter than natural ones, and you will have to run several full cycles per day.

**Rock samples**
Rock samples should be uniform in size, shape and surface area. Rock cores can be made by using a masonry drill and corer, available from hire shops. Alternatively, cubes can be sawn from a block using a tungsten carbide tipped saw. If you do not have either a drill or a saw, you can collect loose rock samples. The disadvantage of this is that variations in size and shape will affect the rate at which they weather. It is not generally a good idea to collect samples using a geological hammer, as this may weaken some samples more than others.

Samples can be obtained from exposed rock sites, but you must take great care, particularly if you are working at the foot of a cliff. Always avoid overhangs.

## Measures of breakdown

The most commonly used measure of rock decay is weight loss, expressed as a percentage of original weight. All fresh samples must be weighed, then weighed again at the end of each weathering cycle once they have returned to room temperature (Fig. 3.2). Results can then be graphed as shown in Figure 3.1.

In the experiments below, times are suggestions only and you may need to increase them if you are using large rock specimens.

**Porosity** is a major determinant of the rate of weathering, so the susceptibility of different rocks to weathering may be explained by variations in the volume of pore space. You should carry out this procedure before you begin the weathering experiments.

1  Weigh the dry rock samples ($m1$).
2  Immerse in de-ionised water for 10 hours.
3  Dry the surface of the samples with blotting paper.
4  Reweigh ($m2$) and express the new weight as a percentage difference of the original weight:
$$\frac{m2 - m1}{m1} \times 100$$

Observation suggests that most sedimentary rocks will be saturated within 10 hours; small samples will have absorbed 80–90 per cent of their capacity within 1 hour.

The method describes the measurement of water uptake which is a surrogate, or representative, measure of porosity. Alternatively, given that the density of water is 1 (kg/litre), you can calculate the volume of pore space provided you know the volume of the rock sample (by displacement): thus,

$$p = \frac{w}{v} \times 100$$

where: $p$ = pore space as a percentage of the volume of the rock
$w$ = mass of water absorbed (kg)
$v$ = external volume of rock sample (litres)

**Heating and cooling** experiments simulate the daily insolation cycles of deserts. Griggs, however, showed that breakdown requires the presence of water, and, while samples should not be cooled in water, they can be soaked when cool.

1  Immerse rock samples in distilled water for one hour.
2  Dry with blotting paper and place in an oven at 60–95 °C for several hours.
3  Cool to room temperature.
4  Weigh and repeat.
5  As a control, leave some samples dry.

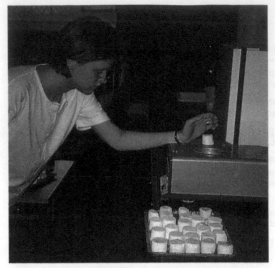

**Figure 3.2** Weighing rock cores to find out the weight loss following a simulated weathering experiment. The cores were obtained by fixing a rock-corer to an electric drill.

**Salt crystallisation** is a physical process. As salts crystallise within pore spaces they exert pressure, which can ultimately lead to rock disintegration. The simplest salt solution is sodium chloride (table salt) but other solutions maybe obtainable from the chemistry department (very effective are $Na_2SO_4$ and $MgSO_4$).

1  Dissolve 10 g of salt in a beaker and immerse the rock sample for one hour.
2  Leave the sample in a warm (not hot) place to dry.
3  Weigh and repeat. The samples may show an initial increase in mass due to the uptake of salt.
4  As a control, immerse some samples in distilled water.

**Freeze–thaw** is a process common in upland Britain. In lowland England and particularly near coastal areas it is not (in present climatic conditions) a significant geomorphological force, as frosts are not all that frequent. Some areas will experience no more than a dozen cycles per year; take this into account when you apply experimental results to field conditions.

1  Immerse samples in water for one hour.
2  Place samples in a freezer.
3  Remove after several hours.
4  Allow to thaw, weigh and repeat.

*Advantages*

The advantage (rare in geography) of these laboratory techniques is that you can control the number of variables and so isolate the particular effect you are investigating. The disadvantage is that, by accelerating the weathering process, you are obtaining results that may be unreliable. With sufficient samples, you can obtain results of statistical significance (see Chapter 14, particularly the Mann–Whitney $U$ test).

## 3.3 Project suggestions

**1** Rates of urban weathering are probably as high today as ever they were because of acidic pollutants from cars and power stations. Observations of the degree of weathering (Tables 3.1 and 3.4) could be related to the geology of the building stone, duration of exposure and location in terms of microclimate (see page 87), and how near pollution sources are. Good experimental design and a number of sampling sites will help you to isolate the relative importance of different factors. For example, the same stone type may be found in buildings of different ages; what evidence of weathering does each show? Were they all exposed to the same degree of weathering or were some more protected? Weathering varies a great deal over even small distances, as Figure 3.3 illustrates.

**2** You can structure a laboratory project around a hypothesis that is developed on the basis of known susceptibilities of different rocks to chemical and mechanical weathering (rock type X decays more quickly than rock type Y). Simulated weathering experiments are designed to test the hypothesis, which is then either accepted or rejected. If the hypothesis is rejected, you should ask questions about how realistic the simulation was. Ideally, the project should be related to local geological and climatic conditions: there is not much point, for instance, in comparing the resistance to weathering of granite and chalk, and by conducting experiments on building stone you can perhaps tie your project into the weathering of buildings.

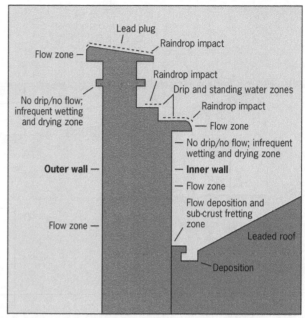

**Figure 3.3 Micro-weathering environments on St Paul's Cathedral**

## 3.4 Slope surveying

The end product of a slope survey is normally a slope profile (Fig. 3.4). A profile is made up of many separate slope units, each of which must be surveyed. Measure the length of each slope unit with a tape and find the slope angle with a clinometer (Fig. 3.6) or Abney level. These instruments can be bought, although details for making a clinometer are shown in Figure 3.5. The method for surveying a slope profile is as follows.

**1** Two people each hold one pole vertically, with the bottom of the pole resting on the ground surface, and with one pole at each end of the slope segment to be measured.

**2** Sight the clinometer from a fixed height on one pole (the top red/white junction on a ranging pole) to the same level on the other pole and read off the angle of slope.

**3** Measure the distance between the two poles. Measurements are usually taken at each break of slope.

**4** The results are then plotted on graph paper to reconstruct the slope profile (Fig. 3.7), bearing in mind that the distance measurements were taken along the slope, not horizontally.

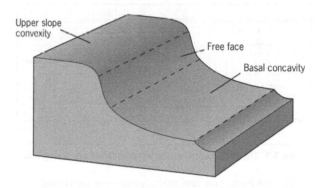

**Figure 3.4 A typical humid temperate convexo-concave slope profile**

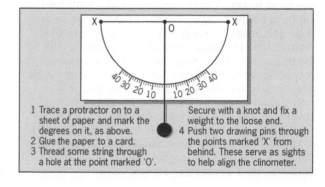

1 Trace a protractor on to a sheet of paper and mark the degrees on it, as above.
2 Glue the paper to a card.
3 Thread some string through a hole at the point marked 'O'.

Secure with a knot and fix a weight to the loose end.
4 Push two drawing pins through the points marked 'X' from behind. These serve as sights to help align the clinometer.

**Figure 3.5 Details for the construction of a clinometer**

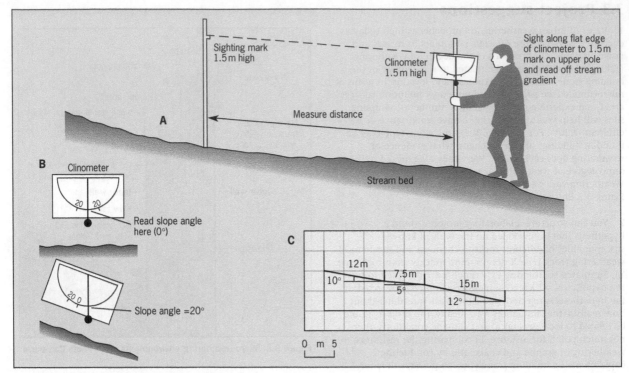

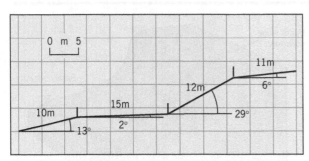

Figure 3.6 Measuring slope profile with a clinometer

**Figure 3.7 The reconstruction of a slope profile from survey data**

Alternatively, you can make your own surveying instrument, a slope pantometer (Fig. 3.8). Two uprights (1 m apart) are bolted loosely together with two cross-pieces. A large protractor and a spirit level are fixed to one upright. Place the pantometer on the slope and, using the spirit level to keep the uprights vertical, note the angle made by the protractor with the upper cross-piece. Step the pantometer downhill perpendicular to the slope contours and note the angle at every step (i.e. every metre). This technique is rather slower than the other one, but it has the advantage that only one person is required.

The method is useful not only for surveying profiles of valley sides, but also for accurate measurement of the shape of other geomorphological features such as drumlins, sand dunes and beaches.

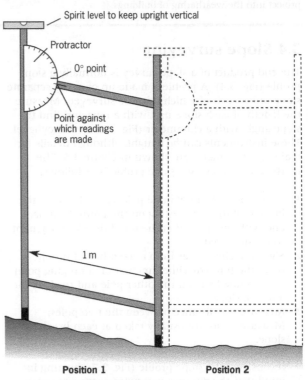

**Figure 3.8 A slope pantometer. The instrument is stepped downhill, and you note the angle made at the upper cross-piece with the protractor at each position.**

## 3.5 Field sketching

A field sketch is a quick hand-drawn summary of a landform or landscape made in the field. An example is shown in Figure 3.9b. The value of a field sketch is that it forces you to look in detail at the features before you. The sketch should be kept simple, bringing out the important and relevant geographical elements of the landscape, and should be labelled.

Turn the paper on its side and draw two lines across the page, dividing the sheet into three. This prevents the common error of too much vertical exaggeration and leaves room for labelling. Using a pencil, draw in the skyline, then a line for the foreground. Fill in the middle, including major features but excluding minor features, other than those (such as trees) that provide scale.

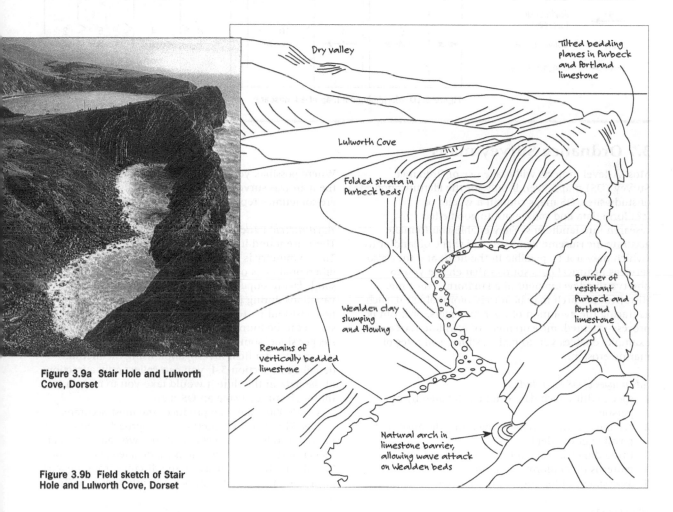

**Figure 3.9a Stair Hole and Lulworth Cove, Dorset**

**Figure 3.9b Field sketch of Stair Hole and Lulworth Cove, Dorset**

## 3.6 Morphological mapping

Morphological maps are maps of slopes and the topography of an area. They are drawn by mapping the terrain using a hand-held GPS to position each mapped element with real world co-ordinates. They can also be drawn by using OS maps and aerial photographs. An example of a morphological map is shown in Figure 3.10. Such maps are useful when studying any geomorphological feature where slope form and slope processes are important, such as slumps, landslides, river terraces and beaches, and in helping to identify and display the grouping of shape components, such as upper convexities, free face and basal concavity, not always made clear by Ordnance Survey maps. They are particularly important for planning the location of settlements and lines of communication.

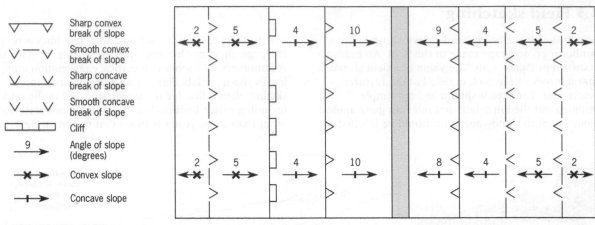

| | | | | | | | |
| Sharp convex break of slope |
| Smooth convex break of slope |
| Sharp concave break of slope |
| Smooth concave break of slope |
| Cliff |
| 9 — Angle of slope (degrees) |
| Convex slope |
| Concave slope |

**Figure 3.10  Morphological map of a landscape**

## 3.7 Ordnance Survey maps

Most A-level projects will make use of Ordnance Survey (OS) maps, even if only for the identification of study sites. OS maps contain a wealth of landform data and can be put to good effect in research into landforms. They enable you to make accurate measurements of size and shape of landforms (which may not be possible in the field) and they also serve as historical data sources that enable you to observe the development of a landform over time. Historical research may be purely morphological, such as tracing the evolution of river meanders, or it can be process-oriented: measurement of coastline retreat, for example, enables you to make estimates of rates of marine erosion.

*Advantages of OS map studies*
1  speed with which data can be generated by one person;
2  accuracy of measurement over long distances not possible in the field;
3  three-dimensional landform reconstructions;
4  patterns of landform evolution;
5  assessment of long-term process rates.

*Disadvantages*
1  cartographic generalisation: the loss of information due to reduction in scale;
2  cartographic inaccuracy, particularly in older maps (which may in any case be small-scale). Older, pre-OS, maps can be used to make generalisations but are usually too unreliable for analysis of any precision;
3  difficulty in obtaining historical maps of a particular locality;
4  scaling problems (comparing old 1-inch maps with modern 1:25 000 maps, for example).

Where possible, you should always refer to the date the map was surveyed rather than published, as maps are sometimes republished using an old survey.

*Asymmetric valleys*
These are found in the chalklands of southern England. The asymmetry is thought to have been caused by the differential rates of mass movement on north- and south-facing slopes as a result of microclimatic variations during periglacial times. Curiously, there is no consistent pattern. Both north- and south-facing slopes can be found (in different places) to be the steeper slope, sometimes within the same valley. If you carry out a field investigation using the method described in Section 3.4, you can collect only a fraction of the data in the time it would take you to make a thorough survey using an OS map.

A 1:25 000 scale map produces the most accurate results. There are two methods of approach. The first is to draw careful cross-sections of the two valley sides at regular intervals and then make a detailed comparison. The alternative is to measure the distance from the axis of the valley to the $n^{th}$ contour on both sides (see Fig. 3.11).

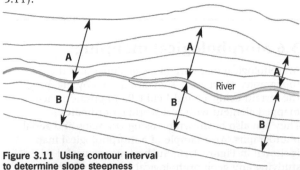

**Figure 3.11  Using contour interval to determine slope steepness**

This gives a surrogate measure of valley steepness (the shorter the distance, the steeper the slope). You can then use a test of significance such as Mann–Whitney $U$ (page 147) to see whether any differences are statistically significant. If you find the two valley sides to be significantly different in steepness, you can carry out further investigation to identify the cause, which could be structural (asymmetric folding) or the result of differing rates of mass movement.

### River meanders

Meander migration is a process described in many books about rivers but, because of the timescale involved, it is seldom possible to observe it. The use of OS maps makes it possible to analyse meander migration, which commonly exhibits two patterns, translation and distortion (Fig. 3.12). Translation is the downstream movement of the meander with its shape unaltered, while distortion describes the process of a

change in meander shape, most dramatically by a cutoff. A survey of the River Axe (Hooke, 1977) showed migration of meanders by up to 80 m between 1843 and 1958 (Fig. 3.13).

An investigation into river meanders could also look at changes in the sinuosity index (length of meandering river divided by the straight-line distance), meander wavelength and amplitude (see page 44).

Only certain types of rivers will display such changes. Rivers crossing an alluvial floodplain are most dynamic, while those that are deeply incised, or bedrock channels, or protected by densely vegetated banks, or artificially restrained, may not move at all.

Non-historical use of maps is also possible. You can relate downstream analysis of meander dimensions and sinuosity to valley topography (and on a large-scale map to channel width), and you can use fieldwork to relate it to downstream increases in discharge (see page 34).

**Figure 3.12  Principal forms of meander change**

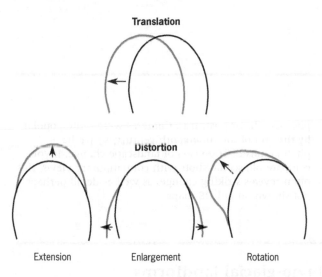

**Figure 3.13  Channel migration of the River Axe, Devon, 1843–1958**

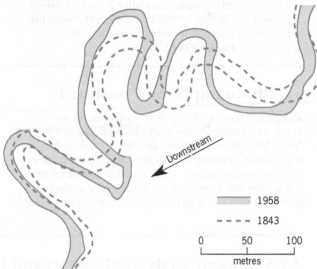

### Coasts

Maps allow accurate measurement of coastal features that may be impossible to survey in the field. The exact planform of a spit, for example, is best taken from an OS map. You can add supplementary detail (such as cross-sectional profile) by surveying in the field (see page 21). You can also add measurements of wave energy and longshore drift (pages 48 and 50), which can be used to help explain the spit's characteristic features (such as alignment and recurves).

Admiralty charts of inshore waters provide an excellent source of information on coastal landforms. Along with OS maps, old photographs and postcards (see Section 3.8), they can be used to trace the development of coastline morphology, in particular coastline advance and retreat.

The east coast of England is especially dynamic: the Holderness coastline, for example, has lost 30 small towns and villages since the Domesday Survey of 1086.

Maps can be used to trace the rate of retreat (or, indeed, the rate of accretion) of the coast. By analysing rates of retreat we can make assumptions concerning the effectiveness of marine erosion. Rates of retreat may vary along a given stretch of coast in response to differing lithologies or because of the planform of the coast, with differential rates of retreat of exposed headlands and sheltered bays. Aggregate rates of retreat of up to 1 m per annum may be recorded. A classic example is the advance of the sea at Mappleton, Humberside, where the cliff has eroded by 1500 m in the last 80 years.

If you have access to a good source of historical maps and a local newspaper archive (County Records Office), then you could add an exciting dimension to this research by investigating magnitude–frequency relationships. Local newspapers usually report particularly damaging storms. Such stories, used together with maps drawn at that time, may make it possible to decide whether coastal retreat was due to a few high-magnitude storms or whether more frequent but lower-magnitude events were more important.

Rates of coastal retreat are generally much lower in the west of Britain, reflecting the more resistant geology, although even here stretches of salt marsh, dunes and spits can be found.

A second use to which you can put maps is a study of place names. These may well give clues that an area of land was once a port, perhaps, or previously much closer to the sea. In Lincolnshire we can find Moulton Seas End, Benington Sea End and Wrangle Bank, all several kilometres inland; Ynys, the Welsh for island, and Ince, an English version of it, may also prove useful clues. Place names can offer a student useful signposts for further research. A comprehensive list, with explanations, can be found in the *Oxford Dictionary of English Place Names*.

*Sources of maps*
Modern OS maps at 1:50 000, 1:25 000, 1:10 000 scales and larger can be obtained from good local bookshops, the Ordnance Survey, or Stanfords, who also supply historical maps.

Admiralty charts can be obtained from the Hydrographic Office at www.ukho.gov.uk. These give an excellent portrayal of the coast; they also show offshore (submarine) contours that are useful for examining patterns of wave energy and movement and the distribution of offshore sediment supply, bars and banks. They are available at various scales up to 1:12 500 and larger, which show superb detail of coastal features. Photocopies of Admiralty charts dating back to 1800 can be obtained from the Hydrographic Office.

County libraries may have a selection of old maps, as may the County Records Office.

## 3.8 Photographs and postcards

Old picture postcards are often available for seaside areas. The postcards are useful because coasts change so fast. Although you can use dated postcards to help you reconstruct geomorphic histories, it is rather a hit-or-miss technique because you cannot guarantee finding the scenes you need; indeed, it may be that such a project begins through the chance discovery of suitable

photographs or postcards. Cameras were quite popular by the end of the nineteenth century, so we have photographs of 100 years of landscape change. If you compare old scenic shots with contemporary views, this often reveals striking change, as well as detail perhaps not shown on old OS maps.

## 3.9 Sediment analysis of glacial and fluvio-glacial landforms

Landforms deposited and moulded by moving ice are generally smaller in scale than major upland erosional features. They are less distinctive in appearance and, because they are made of unconsolidated sediments, they are prone to post-glacial erosion by natural and human agents alike. This means that you cannot always distinguish between hummocky moraine and an esker, for example, on the basis of morphology alone.

If we dig into the unidentified feature, measure and analyse sediment size, shape and fabric (orientation and disposition), we can learn something of the forces that eroded, transported and deposited the material. Through our knowledge of the processes we can be reasonably confident in classifying a landform that might otherwise be impossible to identify.

Sediment analysis is one of the least glamorous

of all geomorphological techniques, but it is an important diagnostic tool for identifying glacial and fluvio-glacial remains. It can also help to explain the mechanics of erosional and depositional processes.

*Sediment size*
This is one of the most useful parameters, but it is potentially one of the most complex. Stones are very irregular in shape, so that measurement of a single axis may prove misleading. On the other hand, measurement of all three dimensions (Fig. 3.14) is difficult and time-consuming. If stone shape is not too highly irregular, the length of the middle axis is most strongly correlated with overall size. It is therefore the most representative single measurement. Fortunately, the middle axis is also the easiest to measure.

a = long axis
b = middle axis
c = short axis

**Figure 3.14 The three dimensions of a stone**

## Pebbleometer

Of the many devices claiming to be pebbleometers, the one illustrated in Figure 3.15, designed by C.R. Thorne and R.D. Hey, is the quickest and easiest to use. It is also the simplest to construct. The pebbleometer consists of a series of ranked squares, each representing a phi value (see below). You have to push pebbles nose first through the smallest hole they will fit through. Read off the phi value and record it. Putting the stone through the hole on its long axis measures the middle axis. A minimum of 25 samples should be measured at each site.

## Phi scale

The phi ($\phi$) scale is an inverse logarithmic scale where a diameter of 1 mm is zero. Table 3.5 gives particle size description and converts the phi scale into millimetres. As most sediments have a log-normal size distribution, a frequency graph of samples on a logarithmic scale (such as phi) produces a normal distribution curve (Fig. 3.16). However, the use of the normal metric scale (measurements in millimetres) is perfectly acceptable. You should use millimetres if you don't understand the phi scale.

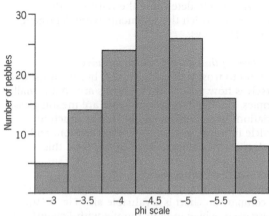

**Figure 3.16 Frequency distribution of sediment size showing normal distribution**

**Figure 3.15 A pebbleometer being used to determine stone size by measuring the middle axis**

**Table 3.5 Comparison of sediment description and units of diameter**

| mm | phi ($\phi$) | Size classes | |
|---|---|---|---|
| more than 256 | more than −8 | Boulder | |
| 256 | −8 | Cobble | |
| 64 | −6 | | |
| 64 | −6 | | Gravel |
| 44 | −5.5 | | |
| 32 | −5 | | |
| 22 | −4.5 | Pebble | |
| 16 | −3.5 | | |
| 10 | −4 | | |
| 4 | −2 | | |
| 4 | −2 | Granule | |
| 2 | −1 | | |
| 2.00 | −1.0 | Very coarse sand | |
| 1.00 | 0.0 | | |
| 1.00 | 0.0 | Coarse sand | |
| 0.50 | 1.0 | | |
| 0.50 | 1.0 | Medium sand | Sand |
| 0.25 | 2.0 | | |
| 0.25 | 2.0 | Fine sand | |
| 0.125 | 3.0 | | |
| 0.125 | 3.0 | Very fine sand | |
| 0.0625 | 4.0 | | |
| 0.0625 | 4.0 | Coarse silt | |
| 0.0312 | 5.0 | | |
| 0.0312 | 5.0 | Medium silt | Silt |
| 0.0156 | 6.0 | | |
| 0.0156 | 6.0 | Fine silt | |
| 0.0039 | 8.0 | | |
| 0.0039 | 8.0 | Clay | Clay |

*Advantages of the pebbleometer*
1 quick to use and therefore generates large sample sizes;
2 gives sizes in phi units rather than the metric scale;
3 samples are grouped into class intervals, which saves time subsequently;
4 readings continue directly from the pebbleometer down to sieve-size material. Particles larger than about 2 mm can be read in the field; smaller matrix material must be measured by using a sieve in the laboratory (see page 63);
5 measures the middle axis, the most important of all.

If you do not have a pebbleometer, you can measure sediments of about 4 mm radius and over in the field with calipers and a ruler.

## Sediment shape

The most useful technique for identifying the nature of the environment in which sediments were deposited is particle roundness and angularity. Sharp, angular sediments are likely to have been deposited directly by ice. Those that have been worn by abrasion in melt-water tend to be less angular and more rounded. A method used to describe particle shape is the Cailleux roundness index.

### Cailleux roundness index
1 Take one pebble from your sample; measure and record the long axis (*a*).

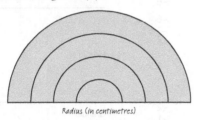

Radius (in centimetres)

**Figure 3.17a Concentric rings for measuring pebble radius for Cailleux roundness analysis. Place the sharpest corner of the stone (in its flattest plane) on the chart to gauge the radius of curvature.**

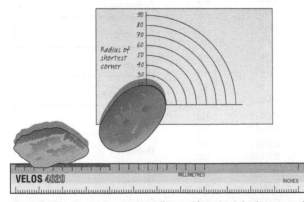

**Figure 3.17b Measuring stone radius and length of the longest axis**

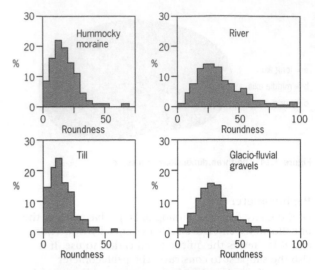

**Figure 3.18 Roundness profiles of sediments from four environments, Wensleydale, North Yorkshire**

2 Holding the pebble flat, lay the sharpest corner on a chart of concentric semicircles of known radii (Fig. 3.17a) and assess the radius of this corner (*r*).

3 Repeat for each pebble in your sample, then substitute your figures in the following equation:

$$R = \frac{2r}{a} \times 1000$$

where *R* = Cailleux roundness index

4 Calculate the mean *R* value for the sample. Values of *R* will lie between 0 and 1000, with 1000 representing a perfectly circular stone.

Results can be plotted on a histogram. The resulting 'profile' will help to determine the nature of the environment in which the sediments were deposited, as Figure 3.18 shows.

*Disadvantages of the Cailleux roundness index*
1 We tend to trust numerical values, but we must ask ourselves how reliable these results are. With small samples, small differences in value are meaningless.
2 Variations in roundness can be due as much to particle lithology as to rounding by erosion, and careless use of Cailleux may fail to reveal this.

### Powers' scale of roundness
For a speedier and more credible technique, Powers' visual comparison chart is a valuable aid (Fig. 3.19). Compare each pebble in your sample with Powers' chart and note the number of particles in each category.

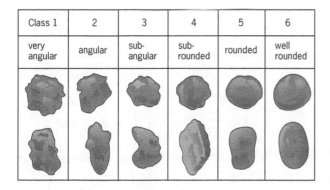

| Class 1 | 2 | 3 | 4 | 5 | 6 |
|---|---|---|---|---|---|
| very angular | angular | sub-angular | sub-rounded | rounded | well rounded |

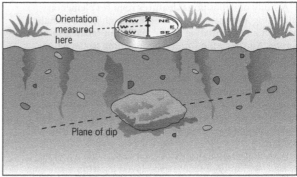

Figure 3.19 Chart for determining Powers' scale of roundness

*Advantages of Powers' visual chart*
1 It is quick to use.
2 The user can ignore micro-structural/lithological controls on sediment shape and so avoid false results.
3 It is better to make subjective judgements with a skilled eye than to end up with misleading numbers.

## Fabric analysis

Fabric analysis refers to the three-dimensional disposition of sediment. You can use the three-dimensional positioning of a particle lodged in a matrix to make inferences about the nature of the depositional process and the direction of the process. You need to take two measurements: orientation (of the long axis on a plan view) and dip (inclination of the long axis on a sectional view). Fabric analysis is used to differentiate boulder clay from solifluction deposits. Both show a preferred (or prevailing) orientation: in the case of solifluction material this is in a down-slope direction, and in the case of boulder clay in the direction of ice flow – which may have been independent of local topography. The dip of solifluction material is roughly parallel with the slope, whereas boulder clay displays no clear pattern.

Depending on the amount of post-glacial disturbance, you may have to dig some way below the topsoil in order to find undisturbed deposits. Vertical faces offer the best access for dip measurements, and horizontal sections for orientations. Only elongated sediments should be used; spherical ones and those smaller than 1 cm should be ignored.

Orientation is measured with a compass and recorded in degrees. It has, by definition, two values 180° apart. When investigating the direction of a depositional process you must record the orientation of the downward-dipping end of the particle (Fig. 3.20).

Dip is measured with a plumbline-and-protractor-type clinometer (Fig. 3.21) and is recorded in degrees from the horizontal.

Plot your data on a circular graph or rose diagram of 180° (dip) or 360° (orientation) – Figure 3.22 – from which you can calculate modal and mean (or preferred) orientations. Use the chi-squared test (page 149) to test the significance of the results before you draw any conclusions.

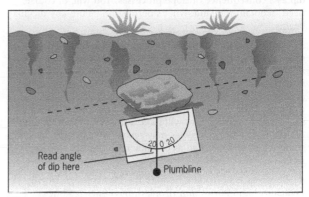

Figure 3.20 Measurement of orientation: the reading is taken at the end of the sediment that is dipping

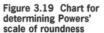

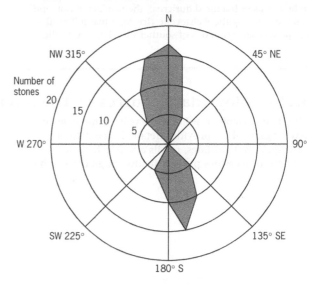

Figure 3.21 Measurement of particle dip

Figure 3.22 Till fabric analysis: the orientation of sediments taken from a Lake District moraine illustrated by a rose diagram

## 3.10 Project suggestions

**1** Lowland glacial deposits are found in a number of British regions. Positive classification of a depositional feature may prove difficult, but it may be possible to identify the depositional environment (glacial or fluvio-glacial).

A study of depositional features requires elementary surveying and mapping. This should highlight the shape, position and orientation of the feature in relation to its setting. You may then find clues pointing to the nature of the processes that led to its formation.

**2** Fabric analysis from a sample of drumlins in the Vale of Eden or Cheshire, for example, could help to establish the direction of local ice flow. It has been said that tools other than a JCB digger are of little value in this exercise, but you could impress the examiner by your careful and thorough method.

Post-depositional processes such as soil-creep and fluvial erosion will undoubtedly have played a part in disturbing the original stratification and dip of sediments. In particular, shallow sediments may have adopted a preferred downslope dip as a consequence of slope processes. You will, of course, need to obtain the landowner's permission before any digging begins.

**3** An investigation into the asymmetry of a chalk valley in southern England could combine the use of OS maps and fieldwork analysis. Formulate a hypothesis such as: *The south slope is shallower than the north slope as a result of accumulation of soliflucted material.* You can test this by excavating and examining the depth of soil and regolith on the two facing sides. Depositional slopes will have a thicker accumulation of debris, and fabric analysis may reveal the existence of solifluction deposits such as coombe rock (flints, chalk and sand).

**4** Some parts of Britain, notably the Marlborough Downs (Wiltshire) and Portesham (Dorset), have sarsen stones, which are hard, siliceous sandstones. These are thought to be blocks of fractured duricrust, the product of hot, arid climates. During the Pleistocene (Ice Age) that followed, the periglacial conditions of southern England led to the

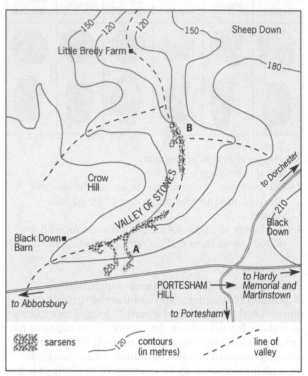

**Figure 3.23 The Valley of Stones showing the sarsen stone streams, Portesham, Dorset**

fracturing and displacement of the sarsens by freeze–thaw and solifluction. Their position and orientation in the chalk dry valleys of the south thus reflect the operation of past processes as well as soil creep under present climatic conditions. Mapping the distribution of the sarsens (Fig. 3.23) and analysing orientation and dip (with appropriate statistical techniques, such as chi-squared, to test for statistical significance) could form the basis of an excellent project.

## 3.11 Geology data sources on the internet

The British Geological Survey has its home page at www.bgs.ac.uk. It provides detailed map data of earthquakes, landslides, fossils and much more.

Some universities publish fieldwork guides on the internet. There are too many to list but typing the place name in a search engine should yield useful results. The Natural England website takes you to Local Geological Sites of interest.

# 4 Hydrology

Hydrology is the study of water and its continual movement through the hydrological cycle (Fig. 4.1). The hydrological cycle can be considered at different scales: global, regional or at the level of the individual drainage basin. Project work is limited to studies at basin or sub-basin scale.

Both the inputs to a basin (amount of water entering as precipitation) and outputs (amount leaving as streamflow or by evaporation and transpiration) can be measured. By comparing inputs and outputs, hydrologists can calculate the net loss or gain of water to a basin over a given time. Any precipitation that does not leave the basin within the period under

observation must be stored within the basin either in lakes, the soil or in aquifers (water-bearing rocks). Similarly, any water leaving that cannot be accounted for by recent rainfall must have been drawn from stores of water in the basin. Even during a very dry summer most rivers continue to flow, fed by water draining from aquifers. However, successive years of drought, particularly in south-east England, and excessive abstraction by water authorities, have made dried-up rivers an increasingly common sight.

The techniques described in this section include most of the methods by which you can investigate the hydrology of a small area.

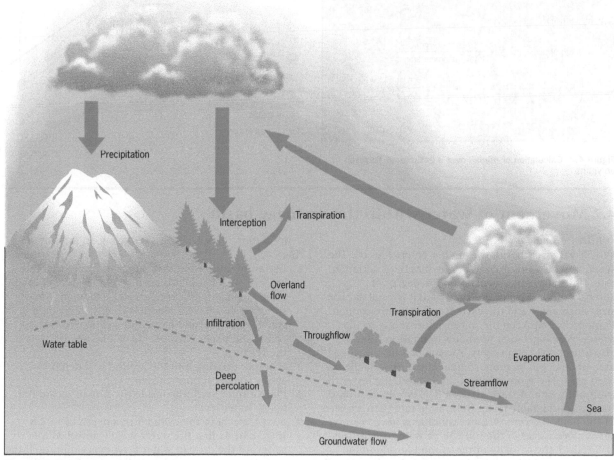

**Figure 4.1  The water cycle**

## 4.1 Inputs

### *Precipitation*

The input of water to a drainage basin in the form of rain is measured with a rain gauge (page 84). However, this records the amount of water falling at one point only. A rain gauge is effectively a point sample, and we assume that the same amount of rain falls over the surrounding area. Over a large area several rain gauges should be used and an average figure for the region calculated using Thiessen polygons (Fig. 4.2).

Not all the rainwater makes its way to a stream, particularly if the ground is densely vegetated or wooded.

If rain lands on trees, it is said to have been *intercepted*; some may be stored on leaves and will eventually evaporate or will reach the ground via the trunk or by dripping straight from the leaves. You can measure the amount of rain intercepted in woodland by placing a rain gauge beneath a tree and another in a nearby open space. The difference in the volumes collected is the interception. Naturally, this will vary according to the type of tree and season. In practice, interception represents a delay or lag in water getting into a stream, as some of it will reach the ground, but much will be evaporated.

---

The standard technique for calculating rainfall over a large area, given the data from a network of rain gauges, is to construct and use Thiessen polygons:

1 Plot location of gauges on a large-scale map.

2 Join gauges with straight construction lines.

3 Bisect construction lines with perpendiculars to produce polygons.

4 With the aid of transparent graph paper, calculate the area of each polygon.

5 Mean rainfall for whole basin

$$= \text{Depth of rain in gauge} \times \frac{\text{Area of polygon}}{\text{Total catchment area}}$$

$$= \left(10 \times \frac{3}{7.5}\right) + \left(15 \times \frac{2}{7.5}\right) + \left(12 \times \frac{2.5}{7.5}\right)$$

$$= 12\,\text{mm}$$

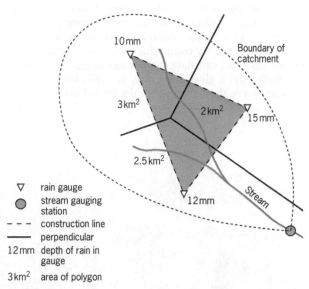

▽ rain gauge
● stream gauging station
– – – construction line
—— perpendicular
12 mm depth of rain in gauge
3 km² area of polygon

**Figure 4.2 Calculation of rainfall over a basin using Thiessen polygons**

---

## 4.2 Movement of water within the drainage basin

### *Infiltration rate*

When water reaches the ground, it enters the soil. The speed at which this occurs is known as the infiltration rate. Infiltration rates depend on the degree of compaction of the ground (how hard it is), soil particle size, slope angle and the current soil moisture content. Plant roots help keep the soil structure loose, so the presence of vegetation promotes infiltration. Land use also affects soil compaction. The soil of a heavily used area of land will be more compact than that of a lightly used area. The more compact the ground, the lower the infiltration rate. A track across a field would accordingly have a considerably lower infiltration rate than a corner of the field. If recent weather conditions have been wet, many pore spaces in the soil will be waterlogged. This will reduce the capacity of the ground for absorbing more water, which will lead to increased surface runoff.

Infiltration rates are measured using an infiltration ring:

1 Use as an infiltration ring a tin can (more than 30 cm tall) with both ends removed, or a piece of tough plastic piping of 15 cm or greater diameter. Make sure that you have a plentiful supply of water, a watch with a second hand and a clipboard with a pencil and paper. Two people will be needed to take the readings.

3 Hammer the ring into the ground to a depth of about 10 cm on your chosen site.

4 Place a ruler vertically inside the ring to record the fall in water level.

5 Pour water into the cylinder to a depth of 15 cm (Fig. 4.3). At first the water will be absorbed quickly, so record the drop in water level every minute. If the rate slows, you may take readings less frequently.

**Figure 4.3 Infiltration rate is measured by timing the speed at which the water level in the tube falls**

**6** As the water level falls, then so too does the water pressure. This alone will reduce the infiltration rate. You must prevent this by ensuring that the ring is kept topped up with water. Add water as quickly as possible and record the amount added on your data recording sheet.

**7** Plot the results on a graph (Fig. 4.4).

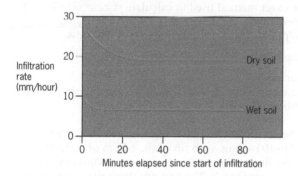

**Figure 4.4 Infiltration curves under contrasting conditions**

## Overland flow

This occurs when the ground is wet and the rain is so intense that the infiltration rate is exceeded. The ground can no longer absorb water, so the excess flows over the surface unchannelled. Overland flow is hard to measure and quite rare in the UK except during heavy storms. The only places you are likely to come across it are at the foot of a slope or near a river channel, and then only under exceptionally wet conditions.

**1** Dig a trench 5–10 m long across the slope, about 30 cm deep and 30 cm wide.

**2** Secure a length of guttering flush with the upslope side of the trench to ensure that all overland flow will be collected. The gutter needs to be at a slight angle to allow the water to drain through piping connected to a measuring cylinder at one end (Fig. 4.5).

**3** Cover the trench with a polythene sheet to keep out the rain.

**4** When overland flow occurs, measure the amount collected in the jug at hourly intervals. At the same time measure rainfall, so that you can get an idea of the conditions that lead to the occurrence of overland flow.

## Throughflow

Water flowing laterally through the soil beneath the surface is called throughflow. It is measured in much the same way as overland flow. The main difference is that the trench needs to be deeper and you need to install a second gutter 50 cm beneath the one at the top of the trench which is measuring overland flow (Fig. 4.6). The first gutter must be retained to prevent overland flow from running into the throughflow gutter.

Note the lag time between rain and the emergence of throughflow. You can relate this to such influences as type and density of vegetation cover, slope angle and to infiltration rates.

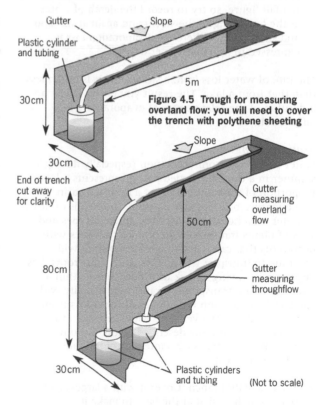

**Figure 4.5 Trough for measuring overland flow: you will need to cover the trench with polythene sheeting**

**Figure 4.6 Trough for measuring throughflow**

*Groundwater*

Water in the soil may eventually drain down vertically into the rock beneath. Here it is known as groundwater, and it may remain stored for months or even years. Volume measurements of groundwater are difficult to obtain, although changes in the height of the water table (the upper surface of the groundwater) can be monitored by observing the level of water in a well. This shows whether the store of groundwater has increased or decreased over a given period. Data on groundwater levels at selected sites in the UK can be found on the Centre for Ecology and Hydrology website (see page 40).

## 4.3 Outputs

Water is lost from drainage basins through evaporation, transpiration and stream flow.

*Evaporation*

Any basin with a lake or other stretch of open water will lose significant amounts of water by evaporation.

You can measure the approximate rate at which water is lost from a lake by using an evaporation pan.

1 Take a large open container such as a plastic washing-up bowl with vertical sides, fill it with water and position it near a lake or pond.
2 Measure the depth of water in the pan at frequent intervals (every few hours on a hot day). Keep it topped up to the original level. If it rains, you must either cover the pan or subtract the recorded rainfall figure, so try to record the depth of water at the beginning of the storm and again at the end and subtract the difference. Evaporation loss is expressed in millimetres per day.

The rate of water loss from a small surface is much less than that from a lake. The result you obtain is just an indication of the relative rate of evaporation on different days.

*Transpiration*

Water transpired from vegetation (especially during summer in a thickly wooded basin) represents a large proportion of the total output of a drainage basin. A single mature oak tree can transpire thousands of litres of water on a hot day. Obviously, different types and sizes of plants transpire at different rates. Trees with deep roots that can reach down for moisture will continue to transpire when grass, with short roots, has begun to die owing to lack of water. Thus, within a drainage basin, transpiration rates vary spatially with changes in vegetation type. Rates of water loss from vegetation also vary over time with changes in the weather. The technique described below, although somewhat crude, will give some idea of the rate at which water is lost.

1 Select a leafy branch. Cover it with a large, clear polythene bag, tied at the base to make it watertight.

2 Use a graduated cylinder to measure the amount of water that has collected inside the bag each day.

The volume recorded can then be related to the day's weather conditions, in particular temperature and sunlight, although of course the bag will create its own microclimate, raising the temperature and protecting the plant from the evaporative effect of wind.

*Stream discharge*

Discharge is one of the more easily measured components of the hydrological cycle. It is defined as the volume of water passing a point on a river bank in a given time (usually one second). The volume is expressed either in cubic metres per second (cumecs) or in litres per second; there are 1000 litres in 1 $m^3$.

**Velocity–area methods**

The exact method used to calculate stream velocity will depend upon the nature of the stream and your equipment. Four methods are described below.

FINDING VELOCITY

*Float method*

In a comparatively clear stretch of river the simple float method will suffice.

1 Select and measure a straightish reach of river 10–30 m long, with no pools, eddies or waterfalls.
2 Use floats whose velocity you can record over the measured reach. The best are those that float mostly beneath the surface and are thus unaffected by wind, for example dog biscuits or oranges.
3 Time the float over the measured distance at least three times and calculate the average. It may be possible to calculate the velocity in the centre of the river and also towards the banks.
4 Multiply the average time by 0.85, because the water on the surface flows faster than under it, and this conversion ensures an accurate velocity reading for the whole cross-section.
5 Calculate the velocity:

$$\text{velocity (metres per second)} = \frac{\text{distance (m)}}{\text{time (seconds)}}$$

See Figure 4.7 for a worked example.

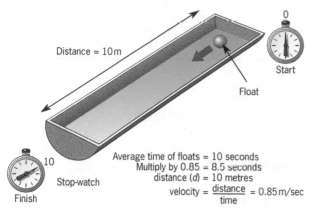

Distance = 10 m

0

Start

Float

10

Stop-watch

Finish

Average time of floats = 10 seconds
Multiply by 0.85 = 8.5 seconds
distance (*d*) = 10 metres

$$\text{velocity} = \frac{\text{distance}}{\text{time}} = 0.85\,\text{m/sec}$$

**Figure 4.7 Measurement of stream velocity**

The disadvantage of this method is the fact that the float (even with the 0.85 constant) does not always flow at the same speed as the water. It can also get caught up in eddy flows and behind boulders and surface debris. If the water itself could be in some way marked, we would have a potentially much more accurate means for measuring its speed. Fortunately such a method exists: the water can be marked with salt and detected with a conductivity meter.

*Salt method*

**Figure 4.8 Measuring the time taken for a wave of salt to pass down a river using a conductivity meter**

A conductivity meter is an instrument used to measure the concentration of solutes in a stream (Fig. 4.8). Attached to the meter is a probe containing two electrodes. When you place it in a stream, solutes (which carry an electrical charge) pass between the two electrodes. The higher the concentration of solutes, the greater the conductivity reading. Conductivity is measured in units called siemens, or micro-siemens (the inverse of resistance units, ohms).

1 Dissolve a handful of salt in half a bucketful of water, and stir to dissolve.
2 Measure a stretch of river of 10–30 m.
3 Empty the bucket as quickly as possible into the stream at the top of the measured reach, and commence timing.

4 At the bottom of the measured reach record the conductivity of the water at 5-second intervals; the salt concentration will pass in a wave on the conductivity meter.
5 Note the time to peak salt concentration.
6 Divide this time by the distance the salt travelled to give time (metres/second).

The advantage of this method over the float method is that you can use it in a boulder-strewn or turbulent stream where a float may prove hopeless. The disadvantages are that you cannot do it by yourself, and you cannot record variations in velocity across a stream's width. The most accurate method of all, and the one employed by professionals, is the current meter.

*Current meter*
Current meters record the velocity at the exact position of the impeller. The position can be adjusted so as to record variations in velocity at different widths and depths across a channel's cross-section.

1 Average velocity is found at 0.6 of stream depth, so hold the impeller at this depth or, if the impeller is adjustable, position it at this depth on the wading rod.
2 Place the impeller in the stream and read off according to the manufacturer's instructions.
3 Record velocity at regular intervals across the channel and either calculate mean velocity or construct an isovel diagram (Fig. 4.9); see page 135 for advice on drawing isoline diagrams.

Current meters tend to be either accurate (but very expensive) or inexpensive and unreliable. Cheap current meters tend to be particularly inaccurate at low velocities, as the impeller needs a high critical stream velocity to make it rotate. Addresses of suppliers of current meters are given in Appendix A6.

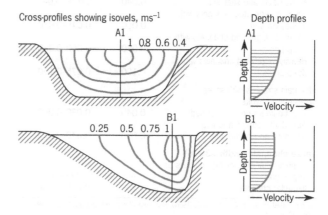

**Figure 4.9 River cross-section showing isovels of stream velocity**

*Manning's formula*

An empirically derived formula expresses stream velocity (*v*) as a function of the three key variables that Manning identified as determinants of channel flow. These are channel gradient (*s*), hydraulic radius (*r*) and a roughness coefficient (*n*). If you measure these three key variables and substitute them in the formula

$$v = \frac{s^{0.5} \times r^{0.67}}{n}$$

you will have an additional method for calculating stream velocity. Alternatively, you can compare observed values (measured using the methods above) against a computed speed.

*Gradient* has to be expressed as a tangent. Simply convert from degrees to tangent using a calculator or tables. Some clinometers have a percentage scale as well as degrees; if so, take readings from the per cent scale and convert to tangent by dividing by 100.

Measurement of the *hydraulic radius* is described on page 43: width and depth measurements must all be in metres. Manning's *n*, the *roughness coefficient*, has to be estimated. It is a measure of the condition of the channel, which includes size and density of bed material and density of channel vegetation, all of which impede the flow. Table 4.1 gives values of *n* for different channel types. Alternatively, by using Strickler's or Kellerhal's formulas, you can get a more objective estimate for Manning's *n*. However, these can only be used in rivers with little weed and lots of pebbles or boulders.

**Table 4.1** Values of Manning's roughness coefficient (*n*) for various types of stream

| Channel type | Normal value | Range |
|---|---|---|
| **Small channels (width <30 m)** | | |
| *Low-gradient streams* | | |
| Unvegetated straight channels at bankfull stage | 0.030 | 0.025–0.033 |
| Unvegetated winding channels with some pools and shallows | 0.040 | 0.033–0.045 |
| Winding vegetated channels with stones on bed | 0.050 | 0.045–0.060 |
| Sluggish vegetated channels with deep pools | 0.070 | 0.050–0.080 |
| Heavily vegetated channels with deep pools | 0.100 | 0.075–0.150 |
| *Mountain streams with steep unvegetated banks* | | |
| Few boulders on channel bed | 0.040 | 0.030–0.050 |
| Abundant cobbles and large boulders on channel bed | 0.050 | 0.040–0.070 |
| **Large channels (width >30 m)** | | |
| Regular channel lacking boulders or vegetation | – | 0.025–0.060 |
| Irregular channel | – | 0.035–0.100 |

*Source:* Based on data in V.T. Chow (ed.) (1964), *Handbook of Applied Hydrology*, McGraw-Hill, New York

**Example**

$$v = \frac{s^{0.5} \times r^{0.67}}{n}$$

if *s* = 2°, tan = 0.034, $s^{0.5}$ = 0.18
if *r* = 0.27, $r^{0.67}$ = 0.42
if *n* = 0.07

$$v = \frac{0.18 \times 0.42}{0.07}$$

*v* = 1.08 m/s

where  *v* = velocity
       *r* = hydraulic radius
       *s* = gradient
       *n* = Manning's roughness coefficient

**Strickler's formula**

$$n = 0.038\, D_{90}^{1/6}$$

where, in a sample of 50 stones, $D_{90}$ is the fifth biggest pebble size.

**Kellerhal's formula**

$$n = 0.038\, D_{90}^{1/4}$$

Which do you find produces the most accurate result?

Having calculated stream velocity, find the cross-sectional area of the stream.

TO CALCULATE CROSS-SECTIONAL AREA
1  Measure the stream width with a measuring tape.
2  Hold the tape taut across the stream and measure the water depth (Fig. 4.10) at regular intervals across the width (e.g. every 25 cm in a 2-metre-wide stream). Take into account the fact that the water will splash up around the rule.
3  Cross-sectional area = *dw*
   where  *d* = average depth
          *w* = channel width
   The average depth is the sum of all the depth readings divided by the number of readings (*n*) plus 1 (Fig. 4.11).

Alternatively (though this is a lengthier process), plot the figures on graph paper to reconstruct the channel cross-section and calculate the area (m²). Discharge can now be calculated by multiplying stream velocity by cross-sectional area (Fig. 4.12). Figure 4.13 is an example of a data sheet that can be used for collecting discharge data.

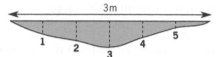

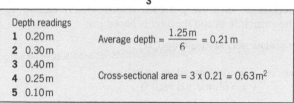

Depth readings
1  0.20m
2  0.30m
3  0.40m
4  0.25m
5  0.10m

Average depth $= \dfrac{1.25\,m}{6} = 0.21\,m$

Cross-sectional area $= 3 \times 0.21 = 0.63\,m^2$

**Figure 4.11  Calculation of average stream depth and cross-sectional area**

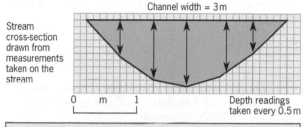

Stream cross-section drawn from measurements taken on the stream

Channel width = 3m

0    m    1

Depth readings taken every 0.5m

Reconstructed cross-section comprises 8 large graph paper squares. Each large square is equivalent to $0.25\,m^2$

$\therefore$ stream cross-sectional area $= 8 \times 0.25 = 2\,m^2$

Discharge $=$ stream velocity $\times$ cross-sectional area
$= 0.85 \times 2 = 1.7\,m^3/sec$ (cumecs)

**Figure 4.12  Calculation of stream discharge using the velocity–area method**

**Figure 4.10  Measurement of width and depth in the East Dart River, Devon**

| **Discharge data sheet** | Date _____ No./Name of reach _____ |
|---|---|

**1  Measurement of velocity**

Length of reach studied _____ metres
Time taken for first float _____ seconds
Time taken for second float _____ seconds
Time taken for third float _____ seconds
Average time of floats _____ seconds
Multiply by 0.85 _____ seconds

Average velocity $= \dfrac{distance}{time} = \dfrac{\underline{\qquad}\ metres}{\underline{\qquad}\ seconds} = \underline{\qquad}$ m/sec

**2  Measurement of cross-sectional area**

Channel width $=$ _____ metres
Channel depth at _____ centimetre intervals = 1 _____   6 _____
2 _____   7 _____
3 _____   8 _____
4 _____   9 _____
Cross-sectional area $=$ _____ $m^2$   5 _____   10 _____

**3  Discharge**

Discharge = velocity x cross-sectional area

$=$ _____ $m^3$/sec

**Figure 4.13  Field sheet for recording discharge data**

## Other methods of finding discharge

SALT DILUTION GAUGING

This method should not be confused with the salt method for measuring stream velocity described above. The principle of the method is based on:

$$c1v1 = c2v2 \text{ (Equation 1)}$$

where: $c$ = concentration (of salt)
$v$ = volume (of water)

$v1$ represents the known volume of water in a bucket and $c1$ the conductivity of salt dissolved in it. $v2$ represents stream discharge and $c2$ represents the conductivity of salt diluted by stream discharge. Solving for $v2$ gives:

$$v2 = \frac{c1v1}{c2} \text{ (Equation 2)}$$

*Method*

1 Fill a bucket with a known volume of water (litres) ($v1$), add several handfuls of salt and stir to ensure complete solution of the salt.
2 Measure the conductivity ($c1$) using a conductivity meter (page 33).
3 Measure the background conductivity of the stream.
4 Quickly pour the saline solution into the stream all at once. It is not necessary to measure the distance between salt injection and sampling site, but the distance must be great enough to ensure complete mixing of the solution with stream water. In a stream with turbulent flow, the distance will be less than in a stream with laminar flow.
5 Read conductivity every 5 seconds as the salt wave passes until conductivity has returned to near normal.
6 Plot a graph of conductivity against time and calculate the number of conductivity units above the background reading. Divide this value by the number of seconds it took the salt wave to pass. This is $c2$.
7 Substitute values in Equation 2 to calculate the discharge in litres per second.

## Example

$$v2 = \frac{c1v1}{c2}$$

$c1$ = conductivity of brine in bucket = 2000 siemens
$v1$ = volume of bucket = 10 litres
$c2$ = mean conductivity of salt wave = 200 siemens
$v2$ = stream discharge (litres per second)

$$v2 = \frac{2000 \times 10}{200}$$

$v2$ = 100 l/s
    or 0.1 cumecs

*Advantages of salt dilution gauging*

1 suitable in turbulent water where measurement of velocity and cross-sectional area may not be possible;
2 suitable in streams that are too deep to wade;
3 with practice it is quicker than velocity–area methods;
4 often more accurate than velocity–area methods.

WEIRS

A weir is a dam across a stream over which water is allowed to flow (Fig. 4.15). Discharge is found by measuring the depth of water above the weir. There are various shapes of weir (of which the most common is the V-notch weir) but the principle of discharge measurement remains the same. All that is needed to calculate discharge is the height of the water above the crest of the weir ($h$). This value is substituted in the formula appropriate to the type of weir (Table 4.2).

**Table 4.2 Formulas for calculating discharge for various forms of V-notch weirs**

| Notch form | Formula |
|---|---|
| 90° V-notch | $Q = 1.38\ H^{2.5}$ |
| 120° V-notch | $Q = 2.47\ H^{2.5}$ |
| 45° V-notch | $Q = 0.69\ H^{2.5}$ |

where $Q$ = Discharge in m³/s
$H$ = Head of water in metres

The method of construction of a weir will depend upon local conditions, but it is unlikely that a channel of more than 50 cm deep can be gauged this way unless help is available. Plywood is probably the best material, and it is necessary to waterproof the weir with plastic sheeting and stones (Fig. 4.16).

*Advantages of a weir*

1 it provides accurate results;
2 once a weir is installed, discharge readings are quick and easy and require only a single person to take readings.

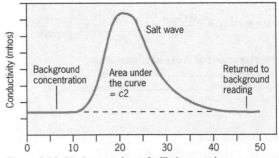

Figure 4.14 Discharge using salt dilution gauging

**Figure 4.15  A V-notch weir**

Stage board

1:2 slope

*h*

**Figure 4.16  A V-notch weir in use on the Findelen Glacier, Zermatt, Switzerland**

*Disadvantages of a weir*
1  it is difficult to construct, install and keep watertight;
2  only suitable on small channels;
3  immobile and therefore of no use in measuring downstream variations in discharge (unless you build a series of weirs).

In weighing up the pros and cons of this method you will need to take into account:
i  the suitability of the site for weir building (including the landowner's permission);
ii  the useful life of the weir or the intensity of its use;
iii  the availability of alternative methods.

STAGE BOARDS
A stage board is a ruler used to measure water level (stage) in a river. Given that stage is directly correlated with discharge, the measurement of stage alone is sometimes sufficient, particularly when, for example, you need only know when peak discharge occurred and not what the actual discharge was. Discharge readings can be obtained from stage boards, provided that you construct a rating curve. The stage board must be anchored firmly to the river bank. Stage is recorded in centimetres.

RATING CURVES
A rating curve is a graph that is used to convert stage (in cm) into discharge (in litres per second or cumecs). The problem is that, because the relationship is unique to each river, you have to derive it by measurement over a range of discharge levels. Rather like using a weir, the rating curve provides quick and easy results once it has been drawn.

*Method*
1  Install a stage board by securing a measuring board or rule firmly to the bank.
2  Using any of the methods described above, measure discharge and stage together over a range of discharge levels.
3  Plot discharge against stage. The relationship is normally logarithmic, so it will produce a straightish line if plotted on log–log paper: this enables reasonably accurate extrapolation, i.e. to read off discharges outside the range of those derived empirically (Fig. 4.17). The rating curve is now complete and ready for use.
4  Read stage level from the stage board and convert to discharge using the rating curve.

*Advantages of rating curves*
1  very quick to measure discharge;
2  discharge can be measured by a single person;
3  an 'organic' method, you can use it as you compile the curve: accuracy improves with additional readings;
4  possible (but with an unknown degree of error) to extrapolate, i.e. to predict discharge at very high and low levels.

*Disadvantages of rating curves*
1  time-consuming to construct: you have to wait for a range of river levels before you can plot many points on your graph;
2  the stage/discharge relationship may be ruined by a flood that deepens or widens the river, in which case the rating curve will have to be redrawn. This can be avoided if you choose a stable bedrock-controlled or boulder-lined reach.

**Figure 4.17 Rating curve showing the relationship between stage and discharge. When plotted on log–log paper it produces a straight-line relationship, so that you can extrapolate or read beyond the end of the line and so predict high and low flow levels.**

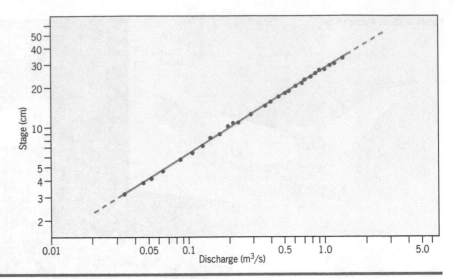

## 4.4 Secondary data sources

The most useful source of information on river flow is the website of the Centre for Ecology and Hydrology National River Flow Archive. Mean daily discharge is recorded in England and Wales at over 1300 gauging stations around Britain. Large rivers are gauged at a number of sites along their length. Data are available in a variety of formats: daily flows, mean monthly flows (with mean monthly rainfall figures), monthly extreme flows, and flow duration statistics. Daily rainfall figures should be obtained from the Meteorological Office (see page 165).

In addition to river discharge, the CEH provides groundwater levels from around 160 boreholes. Figure 4.18 illustrates typical results.

Map showing the location of the Dalton Holme site in northern England

**Figure 4.18 Groundwater levels in the Dalton Holme Estate well**

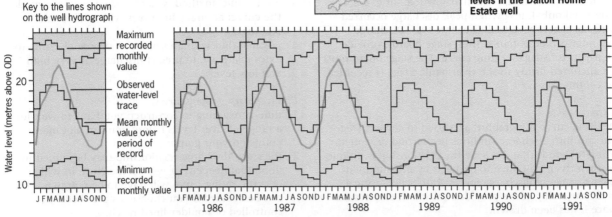

# 4.5 Project suggestions

There are many aspects of drainage basin hydrology that lend themselves to investigation.

**1** Measure infiltration rates in a number of different places and at different times in order to assess the influence of different variables on infiltration. Measure infiltration up a slope to assess the impact of height above a stream. Measure infiltration after both a dry spell and rain to assess the impact of weather. Measure infiltration on different types of rock to assess the impact of geology.

**2** River flooding can be viewed either as a serious hazard or as an interesting hydrological phenomenon. A physical geographer needs to consider questions concerning the susceptibility of a catchment to flooding. Historical data on river flooding can be obtained from the *National River Flow Archive (NRFA)*, using daily river flow. You can then use the data to draw graphs such as a histogram or a flow duration curve (Fig. 4.19). Basin characteristics need to be analysed in order to explain the incidence of flooding. What is the average basin steepness (the difference between the highest and lowest points of the basin divided by its length) and how large is the catchment area? What is the geology? What land uses (including percentage urbanised), or vegetation types, are present? What is the drainage density (km of channel per km²) and what shape is the catchment area?

Of particular interest is the effectiveness of flood prevention measures. Find out what engineering measures or changes in land use have been introduced to reduce the frequency and severity of flooding in recent years. By examining historical flow records, you should he able to see how effective they have been.

Geographers are concerned with the consequences of flooding for people, with local residents' perception of the flood hazard and with the human response. Like many geographical phenomena, floods occur so rarely that you cannot plan a project in the expectation of a flood occurring, so you will almost certainly need to use archive data. Local newspapers are a valuable source of information, and they may include photographs and maps of flood damage as well as descriptions. You can use a questionnaire (page 89) to assess how badly local people have been affected by past floods.

Perception of the *risk* of flooding can also be carried out by using a questionnaire. Studies of natural hazards generally show that people underestimate the risks they are exposed to. A questionnaire might try to find out what residents thought were their chances of being affected by a flood. This could then be related to their location and altitude in relation to the river (in other words, the actual level of risk), the length of time they have lived there, whether they own or rent the house, and the existence of any local flood protection schemes.

**3** Test hypotheses concerned with the relationship between drainage basin variables (such as geology, vegetation, slope gradient, basin size and shape, and land use) and the discharge response. Catchment variables will need to be collected through mapwork or fieldwork. As you will not need to collect the discharge data in the field, you should have time to make a thorough scientific comparison of two basins.

Selection of the basins will depend upon your hypothesis. For example, if your hypothesis was *Discharge in a wooded catchment is less variable than in a non-wooded catchment*, you would have to find one wooded and one non-wooded catchment that were similar in all other respects.

Catchments with a large proportion of built-up area have very different flow characteristics from those that are predominantly rural. Because of the impermeability of buildings and roads, the proportion of runoff is much higher, and the flood risk is greater too. Using an urbanised and a rural catchment in this way, you can observe human influence on hydrology.

You should consider using flow duration statistics (available from NRFA) from which you can draw a flow duration curve (Fig. 4.19). Because this is constructed from a year's data, it largely disguises the effect of individual rainfall events and emphasises the contribution of catchment variables to river regime.

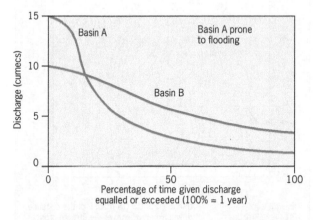

**Figure 4.19 Flow duration curves for rivers with contrasting basin characteristics**

# $5$ River channels

So far, we have considered rivers as part of the hydrological cycle. We have concentrated on measuring the flow of water into, through and out of drainage basins. This chapter describes methods for measuring river channel size and shape, and river load.

## 5.1 River channel form

River morphology is usually considered in three dimensions: long profile, cross-sectional shape and planform.

### Long profile

The long profile of a river can be measured in two ways: the most appropriate method depends upon the size of the river in question and the length of the long profile required. For long distances, the only practical method is the use of a 1:25 000 OS map. A section is drawn using contour lines (Fig. 5. 1).

**Figure 5.2 Measuring stream gradient with a clinometer**

Alternatively, the gradient of a river can be measured with a clinometer. This instrument indicates angles of tilt. You can buy a clinometer or make it (Fig. 3.5, page 21). The measurement of stream gradient is illustrated in Figure 5.2. By recording gradient at regular intervals, you can reconstruct the long profile of a reach (Fig. 5.3).

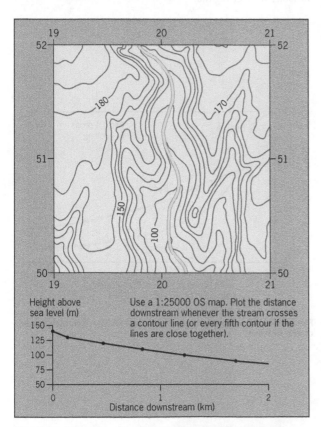

Use a 1:25000 OS map. Plot the distance downstream whenever the stream crosses a contour line (or every fifth contour if the lines are close together).

**Figure 5.1 Measuring river long profile using contours on an OS map**

| Slope measurements | | |
|---|---|---|
| Slope reading | Length of slope measured (m) | Angle of slope ( ° ) |
| 1 | 10 | 13 |
| 2 | 15 | 2 |
| 3 | 11 | 9 |
| 4 | 11 | 6 |

**Figure.5.3 Reconstruction of a channel gradient, using survey data**

## Cross-sectional shape

The cross-sectional form of a river requires measurement of width and depth. Channel width is measured with a rule or tape measure, and channel depth with a rule. Distinguish between channel dimensions and those of the flowing water. In a dry summer, streams may be considerably smaller than their channels. If you are investigating the channel (as opposed to the flowing water), measure the dimensions that would be occupied by water at bankfull discharge, as shown in Figure 5.4. Bankfull discharge is significant, because it is when rivers are at this stage that channel features such as meanders are formed and much erosion of bed and banks occurs.

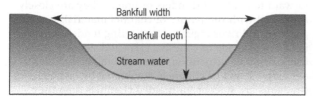

**Figure 5.4  Bankfull channel dimensions**

Cross-sectional form is best expressed in terms of width-to-depth ratio (i.e. bankfull width divided by bankfull depth). The ratio varies according to the composition of the bed and banks: greater than 40 in bedload-dominated gravel channels and less than 10 in suspended sediment dominated alluvial channels.

An important consequence of the width-to-depth ratio is channel efficiency. The efficiency of a channel is controlled by the degree of contact between the bed and banks and the flowing water. Where there is a lot of contact between water and banks (a wide, shallow bedload stream, for example), the loss of energy due to friction is high and the channel is inefficient. Narrower channels with high cohesive banks are more energy-efficient. A measure of channel efficiency is hydraulic radius, which is the ratio of cross-sectional area to wetted perimeter:

$$\text{hydraulic radius} = \frac{\text{CSA}}{\text{WP}}$$

where: CSA = cross-sectional area
WP = wetted perimeter

Wetted perimeter represents the resistance to river flow (friction), and cross-sectional area the energy of the flowing water. Figure 5.5a explains the terms. Hydraulic radius is not expressed in any unit, but the higher the value, the more efficient the channel. Figure 5.5b illustrates the practical difficulties involved in measuring wetted perimeter.

River channels that divide into several sub-channels are described as *braided*. Because water flows in smaller channels rather than in a single large channel, the hydraulic radius tends to be low.

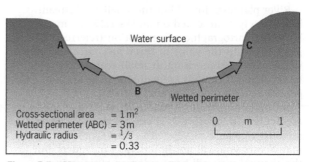

**Figure 5.5a  Measurement of hydraulic radius**

**Figure 5.5b  Measurement of wetted perimeter**

## Planform

River planform (channel pattern) exhibits a range of types (Fig. 5.6). Rivers are rarely straight. A few rivers have distinct meanders and braids, while the vast majority show no regular pattern at all. Geographers can draw upon a wide variety of techniques with which to measure, describe and illustrate river patterns. The guidance below will help you to select the most appropriate technique, although this depends upon what you need the information for.

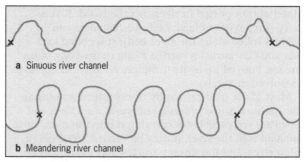

**Figure 5.6  River planforms**

River planform for all but the smallest of streams is accurately represented on an OS 1:25 000 map. However, cartographic generalisation (referred to on page 24) limits the value of this data source in surveys where you need precision in small-scale detail. Nevertheless, OS maps are valuable in terms of a general description of channel pattern.

Meander dimensions (amplitude and wavelength) are defined in Figure 5.7. These dimensions can be measured by field survey with a surveyor's tape. In many cases you will not find perfect meanders, but the channel may still have a degree of sinuosity.

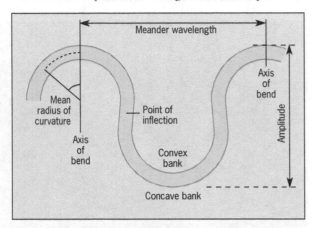

**Figure 5.7  Principal features of meander geometry**

The sinuosity index is used to distinguish channel types. As Figure 5.8 shows, this index is the ratio of channel (or meandering) distance to straight-line distance. A sinuosity index of 1.0 denotes a straight river; greater than 1.5 a meandering river; and above 3.0 a tortuous river.

Braided rivers can be measured by field survey. Here the features to look for are the number (and width) of channels (both dry and water-filled), and the width and length of the braids (the islands or eyots).

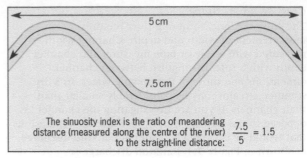

The sinuosity index is the ratio of meandering distance (measured along the centre of the river) to the straight-line distance: $\frac{7.5}{5} = 1.5$

**Figure 5.8  Calculation of the sinuosity index**

Although no universal theory has yet been put forward to explain meanders or braids, they are closely associated with the nature of channel material, stream gradient and river regime. Meandering tends to occur in channels made of cohesive clays and silts with most load carried in suspension, whereas braided patterns are found in gravel channels dominated by bedload. Gradients are steeper for braided channels than they are for meanders, and braided rivers are often found in environments characterised by highly variable discharge regimes. These relationships are summarised in Table 5.1.

**Table 5.1  Classification of stable alluvial channels (*after Schumm, 1977*)**

| Type of load | Bedload (% of total load) | Type of river |
| --- | --- | --- |
| Suspended load | <3 | Suspended-load channel. Width–depth ratio <10; sinuosity >2.0; gradient relatively gentle. |
| Mixed load | 3–11 | Mixed-load channel. Width–depth ratio >10, <40; sinuosity <2.0, >1.3; gradient moderate. Can be braided. |
| Bedload | <11 | Bedload channel. Width–depth ratio >40; sinuosity, <1.3; gradient relatively steep. Can be braided. |

# 5.2 River load

Material transported in rivers is called load. Rivers carry three types of load: bedload (boulders and pebbles rolled along the river bed); suspended load (silt and clay particles carried along in the body of the stream, buoyed up by its turbulence); and solution load (dissolved minerals).

Most of the load carried by rivers originates outside the channel. Slope processes of wash, rainsplash and creep are responsible for carrying it into the river. Once the load is in the river, it may be carried hundreds of kilometres, and the finest particles may eventually be deposited on the sea floor.

## Measurement of bedload

Bedload is carried only at times of high discharge. If flow is swift, then it may be unsafe to wade, so take great care when collecting bedload data. The technique for measuring the volume of bedload involves the use of bedload traps (suitable for small streams only).

1  Dig a trench about 25 cm deep across the stream.
2  Insert a wooden box into the trench. Ensure that the top of the box does not stick up above the river bed. Attach a flap to the upstream side of the box (Fig. 5.9).

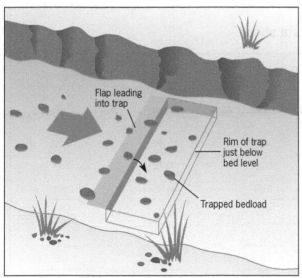

**Figure 5.9  Construction of a bedload trap**

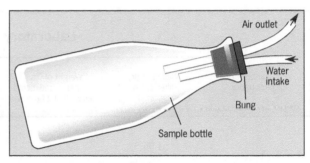

**Figure 5.10  Suspended sediment sampler made from a glass or plastic sample bottle**

3  After a given interval (an hour or a day, depending on the volume of bedload being transported) remove the box from the river and weigh the material.

## Stream competence

As the measurement of total bedload volume is rather an awkward procedure, it may be simpler to determine the competence of a stream. This is the strength of a stream as reflected by the largest-sized particle moved as bedload. Stream competence will change from day to day with changes in velocity and discharge.

1  Take a dozen pebbles of similar size and paint them the same colour with waterproof paint. Take a dozen slightly larger pebbles and paint them a different colour. Altogether you will need about four classes of pebbles covering a range of sizes.
2  Place the pebbles at a marked point in the stream. If the stream is below bankfull stage, only the smallest particles will be moved. When the stream is in flood and flowing swiftly, it may move pebbles of all sizes.
3  Measure and record the distance that the different sizes of pebbles are moved downstream. The colouring will help you to recognise and distinguish the different size categories of your stones. The method can only be used in shallow, clear-running streams.

## Measurement of suspended sediment

There are several types of sampler that can be used to measure suspended load. Samplers are produced commercially, but these are expensive, so you may have to make your own. Use a plastic or a glass flask and piping as shown in Figure 5.10, or simply use a large squash container with a pipe to let out the air.

1  Lower the sediment sampler slowly to the stream bed and then raise it so that the whole depth of the stream has been sampled. Try to judge the speed so that the sampler is just full. Stand downstream of the sampler to avoid stirring up sediment.
2  Measure and record the volume of the whole sample (in litres).
3  If many samples are being collected in the field, allow them to settle before carefully pouring off the clear water.
4  In the laboratory, filter the sediment-laden water through a filter paper of known dry weight. The paper can be placed in a funnel and the water left to drip through. Use Whatman 542 or 541 papers because, although slow, they do filter out even the very finest particles. You can speed up the process by putting the filter paper into a Buchner funnel and connecting it to a vacuum pump (Fig. 5.11).

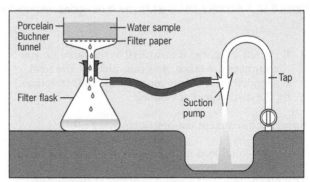

**Figure 5.11  Filtering suspended sediment using a Buchner funnel and suction pump**

5  The filter paper plus sediment is then dried in an oven (105 °C for 2 hours), removed and weighed.
6  By subtracting the known dry weight of the filter paper from the total dry weight, you can find the weight of the sediment sample.
7  As the volume of the original sample has been measured and recorded (step 2), the concentration of suspended sediment can be found and expressed in grams per litre (g/l) (Fig. 5.12).

## Laboratory data sheet

| Number of filter paper | Weight of dry filter paper (grams) | Weight of filter paper + sediment after drying (grams) | Weight of sediment (grams) | Volume of sample (litres) | Concentration (grams/litre) |
|---|---|---|---|---|---|
| 1 | 1.05 | 1.30 | 0.25 | 0.80 | 0.31 |
| 2 | 1.00 | 1.25 | 0.25 | 0.75 | 0.33 |
| 3 | 1.10 | 1.30 | 0.20 | 0.85 | 0.24 |
| 4 | 1.00 | 1.27 | 0.27 | 0.80 | 0.34 |

**Figure 5.12 Laboratory data sheet for use in measuring suspended sediments**

**8** If stream discharge is measured at the same time as the samples are taken, you can calculate the total suspended load: simply multiply the concentration by the discharge. For example:

$$\text{suspended sediment concentration} = 0.2 \text{ g/l}$$
$$\text{discharge} = 25 \text{ l/s}$$
$$\text{total load in suspension} = 25 \times 0.2 = 5 \text{ g/s.}$$

This means that, in 1 second, 5 grams of suspended sediment load passed a fixed point on the bank.

### Measurement of solution load

In some parts of the United Kingdom, particularly limestone areas, most load is dissolved material carried in solution.

A conductivity meter is used to measure the concentration of solution load in a stream. Conductivity is a surrogate (substitute) measure of total dissolved solids. It does not distinguish individual ions (such as calcium and magnesium), but it is the single most useful measure of the concentration of solution load. To convert from conductivity to solute concentration, multiply the reading in micro-siemens by 0.65. This gives an approximate concentration in milligrams per litre.

The conductivity meter enables us to compare the concentration of solutes in different streams. A stream draining a limestone catchment will have a higher concentration of solutes than a stream draining a granite catchment. This is because limestone is more susceptible to chemical weathering. Similarly, the route that water takes through a drainage basin can affect the concentration of solutes. Stormflow (overland flow and throughflow), for example, tends to be low in solutes, because stormflow moves quickly and therefore has little time to dissolve minerals. Groundwater, on the other hand, may have spent months or even years in contact with bedrock and may be almost saturated with soluble minerals. The concentration of solution load (and thus conductivity) will therefore tend to be much greater in groundwater.

# 5.3 Project suggestions

Students undertaking river projects have a number of obstacles to overcome. Access to rivers may not be possible. The size and depth of the river may be too great. Another problem is that rivers change only slowly in a downstream direction. This means that, for studies of hydraulic geometry (the change in a river's dimensions related to increasing discharge), you have to collect data on a scale that is impractical for an A-level student.

1 You can overcome this last problem by comparing two contrasting rivers. The purpose here is to compare rivers displaying widely differing characteristics: meandering and braiding, for example, or bedload and suspended load, but not just variations in size. Both rivers must be essentially alluvial channels (not flowing directly over bedrock, nor bounded by huge boulders).

Measure width, depth, sinuosity and load at 5–10 sites on each of the rivers. If the bedload channel is too deep for a bedload trap, measure bedload size with a pebbleometer (page 27). Channel plan and gradient can be surveyed in the field or taken from a 1:25 000 map. Compare the results of the two rivers with each other and with Table 5.1. Use statistics such as the Mann–Whitney U test to see whether the differences between the rivers are statistically significant.

You must look out for influences on channel shape other than discharge, gradient and load. These influences might be obvious, such as bank strengthening and channelisation (to prevent bank erosion and flooding), the degree of incision of the channel, and protection of the channel by vegetation.

The influences may be less obvious: discharge may be reduced by water abstraction (making the river less powerful and less able to adjust), and there may be changes in land use such as deforestation and construction that lead to increases in sediment load.

2 Another popular and successful idea is a project examining stream velocity. *Stream velocity increases in a downstream direction* is a hypothesis that you can test by observation from source to mouth. The hypothesis has one major drawback: you are unlikely to find a consistent change in stream velocity over a distance of just a few kilometres.

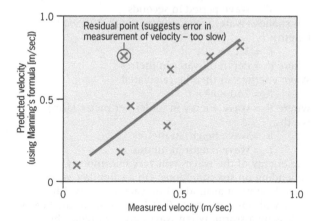

**Figure 5.13 The relationship between directly measured stream velocity and velocity calculated using Manning's formula**

You could compare different methods of measuring stream velocity, or compare Manning's formula with field observation. The question *'How accurate is Manning's formula?'* can be answered by measuring hydraulic radius, stream gradient and channel roughness and comparing the computed values with measured velocities taken at the same site (page 36).

Plot a scattergraph of observed results against expected results (Manning). If Manning matched your observed results exactly, your graph would show a 45° diagonal line (Fig. 5.13). If your observed and expected results are not alike, how can you know which set of data is right and which is wrong? Consistency is one answer to that question: if one set of readings shows wide fluctuations, it is the one most likely to be unreliable.

3 Page 25 contains suggestions for a project that uses old maps to trace the development of river meanders. A project to investigate present-day meander patterns might test the hypothesis that *Channel width is related to meander wavelength* Gregory and Walling (1973) quote a ratio of between 1:6 and 1:10. In order to generate sufficient data, it may be necessary to measure dimensions from OS maps. The major problem with this is that small inaccuracies of measurement may result in considerable over- or under-exaggeration of the actual channel size. Data can be collected in the field but, as explained earlier, unless you have access to several suitable rivers of different sizes, you will not find the variation you are looking for.

4 Using a 1:25 000 OS map, test the hypothesis that *Sinuosity increases with gradient*. You will need to measure the gradient and sinuosity of about 50 1-km reaches and plot the results on a scattergraph. You can measure the strength of association using the Spearman rank correlation coefficient (page 144) and test for significance using the chi-squared test (observed sinuosity compared with expected sinuosity; the null hypothesis predicts no variation in sinuosity, and therefore your expected value equals the mean sinuosity of all 50 reaches, page 147). You will need to refer to a fluvial geomorphology text to look for explanations, if you accept the hypothesis.

5 Load responds to changes in discharge: test the hypotheses, *Suspended sediment load increases with discharge* (the erosive and wash effect of rain), or *Concentration of solution load decreases with increases in discharge* (the dilution effect of stormwater). Such projects require measurement of load over a range of discharges (although stage measurement will do), and you will need to plot both on a time-series graph. Which of the two hypotheses you choose (sediment or solute) will, no doubt, depend on the geology of the area you live in.

Sediment yield is strongly influenced by vegetation cover and land use. Careful experimental design may enable you to monitor the differential sediment yields of catchments. You may be able to relate this to differences in percentage of bare earth, or type and density of vegetation (including agricultural land use), and soil erosion accelerated through mismanagement of land.

# 6 Coasts

The UK provides many opportunities for studies of the coast, for it has an extraordinary variety of coastal environments and superb scenery. The coast is also one of the most dynamic natural environments, and in many places you can see change take place in just a matter of months. The coast is unusual in the extent to which it is the product of all three geomorphological elements: rock, water and air. It is also the focus of great human pressure: industry, housing and recreation all take their toll, and in no other environment do people try so hard to control the destructive forces of nature.

## 6.1 Wave energy

One of the most destructive of all geomorphological processes is wave action. Upon this depends the balance of erosion or deposition. The rate of accretion or erosion will also, in part, be determined by the ferocity of wave attack. Coastlines are described as high- or low-energy environments according to the energy of incoming waves. Wave energy is controlled by the size and frequency of waves and is measured by using the following technique.

**Wave height** (see Figure 6.1 for an explanation of wave dimensions) is measured (when possible) by getting into the water or standing on a groyne with a ranging pole. See how far up the pole the wave crests reach and how low down the pole the wave troughs dip. The distance between crest and trough is the wave height. Alternatively, you may be able to observe waves passing a fixed obstacle such as a pier leg or groyne. In this case, estimate the average height of the waves against the object.

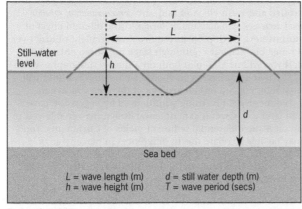

**Figure 6.1  Wave geometry**

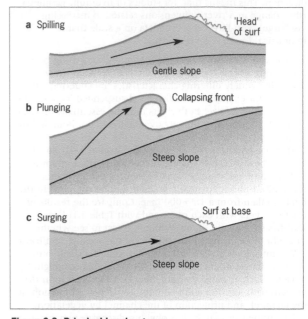

**Figure 6.2  Principal breaker types**

**Wave length** is more difficult to assess by eye. Fortunately, wave length is related to wave period (the time between two wave crests passing the same point). Time fifty waves passing a fixed point and calculate the average wave period. Wave length can then be found from one of the following equations:

For deep-water waves (water depth > $^1/_2$ wave length):
$$L = 1.56 \times T^2$$
where $L$ = wave length in metres
$T$ = wave period in seconds

For shallow-water waves (water depth < $^1/_2$ wave length):
$$L = 3.13 \times T \times \sqrt{d}$$
where $d$ = depth of water in metres

Wave energy can then be calculated:
$$E = 740 \times h^2 \times L$$
where $E$ = wave energy in joules per metre (J/m) width
$h$ = wave height in metres
$L$ = wave length in metres

The energy of the waves will vary enormously, depending on sea conditions. On a calm day, with wave heights of around 0.5 m, each metre width of wave will transfer about 1000 J of energy to the beach. Following a storm, when waves are 3 m high, the energy expended by them will be over 1 000 000 J/m width.

Since 1 watt = 1 joule per second, you can divide joules per metre by average wave period to express wave energy in watts. This gives you some idea of wave energy in terms of light-bulb power.

Coastal geomorphologists describe waves as one of three types, illustrated in Figure 6.2. A problem arises when the waves under observation fail to fit neatly into one of the three categories. Phase difference is a measurement that enables you to categorise waves quite easily.

*Phase difference*
Phase difference requires two measurements: swash time and wave period. Swash time is the time elapsed, in seconds, between the breaking of the wave and the swash reaching its highest point up the beach. Wave period (referred to above) is the average time between wave crests. Time 11 waves and divide by 10.

$$\text{phase difference} = \frac{\text{swash time}}{\text{wave period}}$$

The phase difference of surging breakers is around 0.5 and of plunging breakers less than 1.0. Spilling breakers have a phase difference greater than 1.0.

As an alternative to phase difference, waves can be categorised according to their steepness.

**Wave steepness** is defined thus:

$$\text{wave steepness} = \frac{\text{wave height}}{\text{wave length}}$$

Values for wave height and wave length must, however, be collected in deep water (i.e. where water depth is greater than wave length). The practical application of wave steepness is its use in categorising wave types. The significance of wave type for studying erosion is considered below.

# 6.2 Beach profiles

A beach profile, or cross-section (Fig. 6.4), reflects several factors. The most important are the size and permeability of beach material and the steepness and type of wave that modelled it. Generally, beaches made up of shingle can support quite steep gradients, around 1:4. The Abbotsbury end of Chesil Beach in Dorset, which is composed of duck-egg-sized stones, maintains gradients of between 1:2 and 1:3. Fine sandy beaches support gradients not usually greater than 1:30. Larger beach sediments are more stable and have higher angles of repose (Table 6.1). In addition, the high permeability of large stones means that on such beaches there is a net upbeach sediment transfer, because the power of the backwash is diminished (Fig. 6.3).

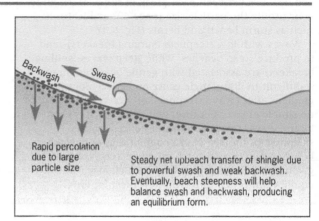

Figure 6.3 The effect of sediment size on swash and backwash, and beach slope

| Table 6.1 Beach gradient and particle size *(Source:* Clowes and Comfort, 1987) | |
|---|---|
| *Particle size* | *Beach gradient* |
| Cobbles | 24° |
| Pebbles | 17° |
| Granules | 11° |
| Very coarse sand | 9° |
| Coarse sand | 7° |
| Medium sand | 5° |
| Fine sand | 3° |
| Very fine sand | 1° |

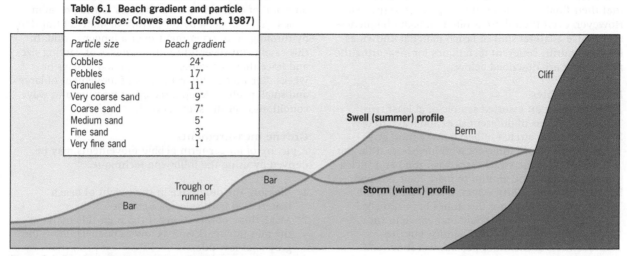

Figure 6.4 Beach profile illustrating typical summer and winter forms

You can measure percolation rates on beaches by pushing an open tube into the beach, filling it with water and recording the time taken for the water to disappear. This time is, of course, much shorter on shingle than on sandy beaches, i.e. there is an inverse correlation between particle size and percolation rate. A simple but quite effective method for measuring the balance of forces of swash and backwash is the swingometer, which is available commercially or can be made (Fig. 6.5). You push the swingometer into the sand or shingle in the swash/backwash zone and estimate the relative strength of swash and backwash by how far the arm is tilted by the two forces. The balance of forces will reflect the steepness of the beach and percolation rate.

The method for profiling beaches is exactly the same as is described on page 42. With a clinometer, two ranging poles and measuring tape, record the gradient and distance between each break of slope. Mark on to your profile the particle size (page 49) and features such as storm beach and berms (Fig. 6.4).

Waves with low steepness (surging breakers) tend to produce steep beaches, while steep waves (spilling breakers) are associated with gentle beach gradients. It is difficult to disentangle cause and effect: does beach gradient affect wave steepness, or vice versa? There is no simple answer because, while the beach is shaped by the waves, wave form is influenced by the steepness of the beach. An equilibrium, or balance, between waves and beach might be achieved but, given the speed at which they both change, the equilibrium beach form will never last for long.

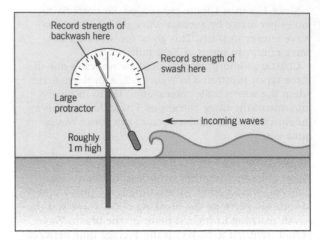

**Figure 6.5 Swingometer for measuring relative strength of swash and backwash**

Geographers agree, however, that there is a connection between wave energy and beach steepness. Beaches commonly have a summer and a winter profile. The summer profile, gently shelving and usually sandy, is shown in Figure 6.4, as is the steeper, stepped and often shingle-dominated winter profile. High-energy storm waves throw shingle high up the beach beyond the swash/backwash zone and cut the sandy forebeach, depositing the sand offshore. Sorting of sediment by size (see Section 6.4) and tidal variations account for the steps or berms. Lower-energy summer waves fill the cut profile with sand and produce the gentle beach gradient.

# 6.3 Longshore drift

Waves often arrive oblique (at an angle) to the beach. Wave refraction causes them to be bent around, so that their final approach is nearly parallel to the beach. However, even then there is often enough obliqueness to produce a lateral shift of beach material known as longshore drift. Different techniques for measuring the rate of drift are described below.

### Marker pebbles

1  Spray-paint an assorted sample of at least two dozen pebbles of different sizes and leave to dry.
2  Select a clear stretch of beach away from any groynes that will interfere with longshore drift. Put the stones in the swash/backwash zone and stick a pole firmly in the ground to mark their position.
3  After a count of fifty waves, locate and plot the direction and distance the stones have travelled. Some will vanish and some may move in the opposite direction to the majority, but the movement of the other pebbles will show you the direction and rate of drift.

This experiment tells you the direction of longshore drift at a particular moment. However, because the direction of drift varies with changes in the prevailing winds, you will get a better picture of material transfers by running your experiment over a longer period. In this case, paint several hundred pebbles of varying sizes and trace their movement over a period of days or weeks. Record the different rates of movement of large and small pebbles. Keep a record of weather and wave conditions over the same period.

### Groyne measurements

A picture of longer-term pebble movements may be derived by using the following technique.

1  Measure the difference in the height of beach material on both sides of a groyne.
2  Make these observations at the high-tide mark, at the mid-tide mark and as low as possible down the groyne at low tide.
3  Plot your results as shown in Figure 6.6.

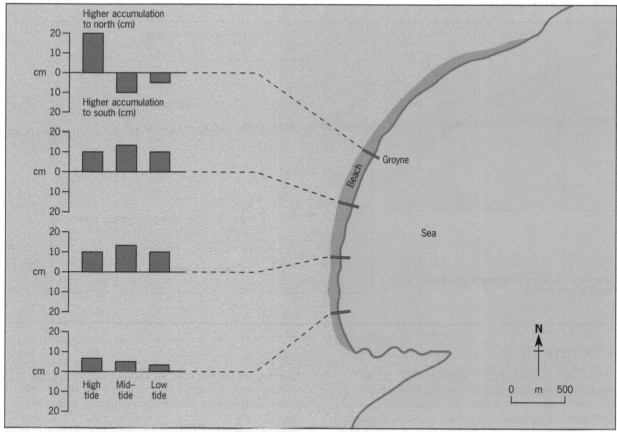

**Figure 6.6  Medium-term patterns of longshore sediment movement on Swanage beach, as revealed by the accumulation of sand and pebbles against beach groynes**

## 6.4 Beach sediment analysis

Geographers can learn much about the processes that shape coastal features by examining the size and shape of particles along the shoreline. For example, the roundness of a pebble shows the amount of attrition, or wearing, to which it has been subjected. Freshly shattered fragments from cliffs are sharp and angular, but after prolonged wearing by waves they become more rounded and smaller. On a beach with varied geology, we normally find a correlation between distance from the outcrop and roundness of the pebbles from that source. Sediment shape and roundness are dealt with on page 28.

**Sorting**

Casual observation of beach material will often reveal pebbles ordered by size at intervals up a beach. This reflects the sorting action by waves and possibly also the availability of different-sized pebbles. Measurement of particle size is covered in Section 3.9.

Waves of a given energy can shift particles of up to a certain threshold size. Owing to percolation, swash is more powerful than backwash. This means that stones of all sizes are thrown up the beach, but only the

smaller ones are dragged back with the backwash. Variations in wave energy from day to day, and changing tide levels, superimpose new patterns to produce a complex and dynamic picture.

A given section of beach often does not contain the full range of particle sizes. One explanation is lithology: for example, chalk particles less than 2 cm in diameter are rare, because they are easily broken down when they are so small. Analysis of sediment size reveals modal (normal or skewed) or bimodal size distributions.

Lateral sorting of beach material is also common. Chesil Beach, referred to earlier, displays a steady increase in stone size from west to east, as can be seen from the result of an A-level survey by Rupert Garton in Figure 6.7. This sorting has been attributed to two causes. Firstly, large particles tend to travel faster and further than small ones (which get trapped in the interstices). Secondly, the greater energy of waves from the west (due to the longer fetch) moves stones of all sizes in an easterly direction, but the weaker waves from the east can shunt only the smaller particles back.

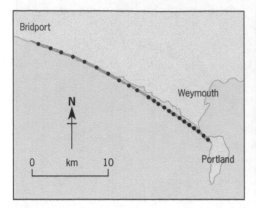

Figure 6.7 **Variation in sediment size along Chesil Beach, Dorset**

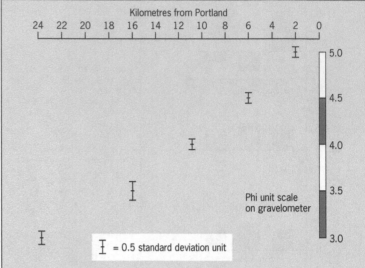

# 6.5 Cliffs

Cliff studies involve the comparison of profile and planform with geological factors and the intensity of wave attack. Coastal cliffs are perhaps the most dangerous environment in which to work. If you are working at the foot of a cliff, you should wear a hard hat and keep well clear if there is any evidence of slope instability or falling stones.

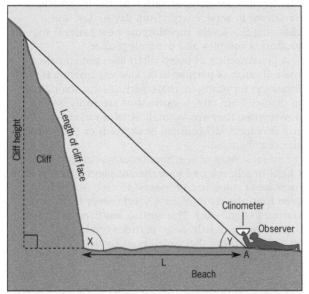

Figure 6.8 **Measuring cliff height using elementary surveying and trigonometry**

*Cliff profile*

Cliff height is either read off an OS map, estimated visually using a ranging pole as a scale or, better still, determined by elementary surveying (Fig. 6.8). Cliff slope is measured with a clinometer (page 21) and its geology identified, if necessary, with the aid of a local guidebook or geological map (page 17).

The steepness and profile of a cliff reflect the balance of the forces against it and its own internal resistance to the beating that it receives. The balance of sub-aerial weathering of the cliff face against basal erosion of weathered material helps to determine cliff steepness. Cliffs that experience active wave undercutting are steeper than dead cliffs (see page 57) that have basal protection, whether in the form of a spit, salt marsh, promenade, a raised beach or an accumulation of dumped material.

The geological factors that determine the form of the cliff are structure and lithology. Major structural controls include faulting and the angle of dip of the rock, which affect the overall profile. Minor structures, such as jointing and spacing of bedding planes (i.e. the thickness of the beds), are more important in affecting resistance to erosion. Lithology, or rock type, influences the susceptibility of the rock to sub-aerial weathering, and sheer rock hardness is a major factor determining equilibrium slope steepness and the presence, or absence, of cliff overhangs.

A study of cliff form must include observation and measurement (where possible) of all these geological variables. You must also look out for evidence of slope processes: basal erosion, mass movement and weathering.

*Cliff plan*

The planform of a cliffed coastline is more strongly influenced by geology than by wave attack. However, as any geomorphological textbook will explain, wave refraction and the degree of exposure of a coast to prevailing winds and marine erosion do play a significant part. This is illustrated for Swanage Bay in Figure 6.9.

One approach for a study of a cliffed coastline is to examine the relative importance of the factors that influence the plan of the coastline at different geographical scales. At the regional scale, OS and geological mapwork will show the relationship between the lie of the rocks and the broad run of the coast. On a smaller scale, rock lithology and relative rock hardness may be shown to determine the pattern of headlands and bays. At an even smaller scale, fieldwork will reveal that, within a given lithology, minor structural variations are responsible for the smallest-scale expressions of the coast's planform.

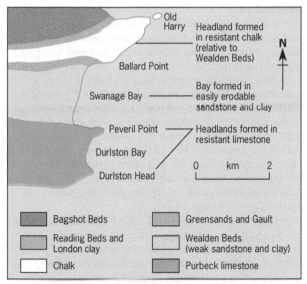

Figure 6.9 Swanage Bay, Dorset, illustrating the relationship between rock type, structure and landforms

## 6.6 Sand dunes

Studies of sand dunes usually focus on the biogeography of such areas. The relevant techniques are found in Sections 7.1 and 7.3, Soils and Vegetation respectively.

Sand dunes are accumulations of wind-blown sand. They occur only in areas with vast supplies of sand, either in the form of a spit or a wide beach, normally with a high tidal range. While the source of sand will vary from one set of dunes to another, the sand in many of Britain's dunes is thought to have been a one-off injection supplied by rising seas during the Flandrian marine transgression (10 000–7000 years ago). This view suggests that dunes (and indeed most beaches) are ephemeral (short-lived) features which need conservation.

Dune studies are normally based on transects taken inland from the top of the beach (or fore-beach). The newest dunes (embryo dunes) develop at the top of the beach, and the oldest dunes are found furthest from the sea. This approach allows you to study the evolution of dune morphology, soils and flora over time.

### Dune morphology

A transect running from the beach inland reveals a typical sequence of dune ridges and depressions (slacks), shown in Figure 6.10. Elementary surveying with ranging poles, measuring tape and clinometer (page 21) will reveal the exact pattern. Dunes grow in size (with age) as sand accumulates (deposition exceeds erosion) usually until dune 1. The dunes then get progressively smaller inland as the supply of sand is cut off and erosion exceeds deposition.

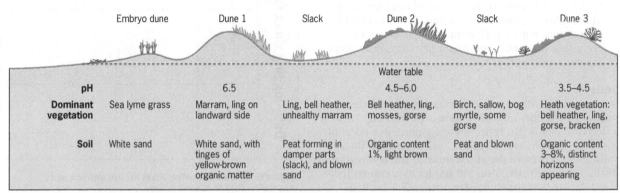

| | Embryo dune | Dune 1 | Slack | Dune 2 | Slack | Dune 3 |
|---|---|---|---|---|---|---|
| pH | | 6.5 | | 4.5–6.0 | | 3.5–4.5 |
| Dominant vegetation | Sea lyme grass | Marram, ling on landward side | Ling, bell heather, unhealthy marram | Bell heather, ling, mosses, gorse | Birch, sallow, bog myrtle, some gorse | Heath vegetation: bell heather, ling, gorse, bracken |
| Soil | White sand | White sand, with tinges of yellow-brown organic matter | Peat forming in damper parts (slack), and blown sand | Organic content 1%, light brown | Peat and blown sand | Organic content 3–8%, distinct horizons appearing |

Figure 6.10 Cross-section of sand dunes on South Haven Peninsula, Dorset, showing soil and vegetation characteristics

## Vegetation

Without vegetation, dune morphology would undoubtedly be very different. Plants provide shelter for the deposition of wind-blown sand, they trap sand and anchor the dunes.

Plant succession across a dune transect is called a psammosere. The embryo dune and dune 1 zone of net accretion is a hostile environment. The organic content of the soil varies from tiny traces to zero; soil salinity is high and the dune is frequently soaked by sea water. Plants are sometimes buried under a metre of sand each year; drought is frequent and often prolonged; and winter storms wreak havoc. Only a few highly adapted plants (sea lyme grass, sea couch grass and, above all, marram grass) flourish here.

A short distance further inland, conditions become less hostile. Soils are grey with accumulated organic matter; dunes are no longer inundated by sand and sea water; the water table is higher and conditions are more sheltered. Not surprisingly, a greater variety of plants flourishes. This biological richness is due to the greater opportunity for colonisation by pioneer species during the soil's longer history. The oldest dunes may support a climax vegetation that will often be a heath assemblage.

A distinctive feature of dune vegetation is the marked variety in the composition of plant communities between the ridges and slacks. This is because the shelter offered by the slacks is greater, the water table is closer to the surface and the soil is less acid (a consequence of the accumulation of bases leached from the ridges).

The techniques you use will depend upon the aim of your survey, but a summary of the techniques you might consider using includes:

1 plant enumeration: the number of different plant species within a quadrat;
2 percentage of ground cover (or its inverse, percentage of bare earth);
3 identification of the dominant plant species, its proportion of ground cover and the appearance of new pioneer species;
4 plant condition: measurement of height of grasses and trees; chest-height circumference of trees and general appearance of plants. See Section 7.3 for details of techniques. Figure 6.11 is an example of a data recording sheet that you might like to copy or adapt for your own purposes.

## Soils

Beneath the plants, less spectacular but nevertheless important changes take place along the transect. There is a decrease in soil pH, reflecting the cumulative effect of leaching; an increase in organic matter content; thicker soils; and eventually (by about dune 2) the emergence of distinct soil horizons. Also, but harder to measure, there is a decrease in soil salinity. See Section 7.1 for details of appropriate techniques.

**Data-logging sheet for use on a sand dune**  Date _____

| Location | % ground cover | No. of species | Dominant species | Condition (height, health, % cover) | Secondary species | Condition | Soil depth | Soil colour (organic matter) | Soil pH |
|---|---|---|---|---|---|---|---|---|---|
| | | | | | | | | | |

Figure 6.11  Data-recording sheet for use with soil and vegetation studies in sand dunes

## Dune erosion

The balance between accretion and loss of sand is tilted in favour of erosion if vegetation is destroyed. This is not an uncommon event in dunes: burrowing and grazing by rabbits, fire, trampling, and motor-bike damage all play a part. Once sand is exposed, it erodes quickly, producing a blowout or a depression in the ridge.

It is not always possible to attribute a blowout to any one cause, as often the evidence is dispersed along with the sand. Nevertheless, it is worth noting the pattern of footpaths and the difference in vegetation cover in heavily used areas and areas that are out of bounds. Natural England and the National Trust sometimes fence off areas to protect the marram grass, and the result is quite striking. In many areas wooden duckboards have been put down, and people are encouraged to stay on the paths. A project suggestion for investigating footpath erosion is found on page 125.

# 6.7 Salt marshes

Salt marshes form in sheltered estuarine locations. Such marshes grow where there is a large tidal range. Mudflats develop in quiet water, the product of sediment brought up on the flood tide and silt deposited by the river. Colonisation of mudflats by halophytic (salt-loving) mats of algae, *Zostera* (eel grass) and *Spartina* grass boosts the rate of accumulation and the vertical growth of the emerging salt marsh. Environmental conditions change with continued growth, as the marsh is submerged less frequently and for shorter periods. Colonisation by *Juncus maritimus* and freshwater reeds and sedges completes the plant succession, which is called a halosere (Fig. 6.12). Marshes are a potentially dangerous environment in which to operate, and you should not work there unaccompanied. You must be aware of the dangers of mud and quicksand, and the speed of the incoming tide.

There is little of morphological interest except the appearance of a creek system at a late stage of marsh development. Studies of salt marshes therefore tend to concentrate on plant succession. When sampling soils or vegetation on a salt marsh, construct a profile using the surveying techniques described in Section 3.4. Rather than sampling at regular intervals inland from the sea, collect your samples at regular intervals (25 cm) above the low-water mark, i.e. regular altitudes, since the key influence on plant habitat is duration and frequency of inundation by the sea. Having taken your own transects in the field, compare the results with a model such as Figure 6.12. Techniques for measuring and analysing salt marsh soils and vegetation are described in Sections 7.1 and 7.3.

In the twentieth century, the salt marshes of eastern England in particular have come under increasing attack from the sea. The tectonic tilting of the British Isles to the south-east, together with rising sea levels, has led to a serious loss of salt marsh. Many marshes are trapped in front of sea walls. They cannot move inland in response to higher seas and are therefore being eroded away. However, salt marshes are now recognised as being one of the most cost-effective forms of sea defence, and Natural England is continually generating new salt marsh, mostly by allowing old sea walls to be demolished by the sea.

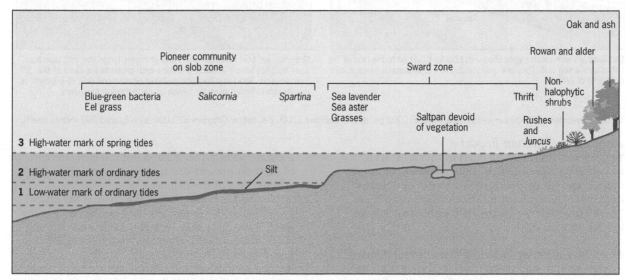

**Rgure 6.12  Sketch of plant succession in a halosere, Llanhridian Marsh, Gower Peninsula, South Wales**

## 6.8 Coastal protection

For centuries people have been drawn to the coast, whether to trade, to fish, or for recreation and relaxation. The supporting towns, harbours and resorts are all exposed to the risk of erosion or deposition, and from early times people used engineering measures to limit the damage. As people's mechanical power and

ambition have grown, human impact has become more effective and (in roughly equal proportion) less wisely applied. Figure 6.13 illustrates some of the more commonly used forms of coastal protection with some idea of their role and cost.

**Figure 6.13 Methods of coastal protection**

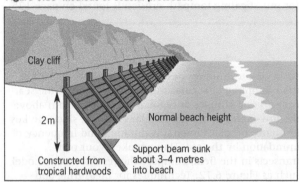

**Revetments** are used where the expense of a sea wall cannot be justified. They break the force of the waves and trap beach material behind them to protect the base of the cliffs. But they do not give total protection to the cliff foot as a sea wall does.

**Sea walls** are the most effective means of preventing erosion but they are also the most expensive. They deflect the power of the waves, but this usually means that the waves wash away the beach material and can undermine the sea wall. Groynes are needed to hold the beach in place.

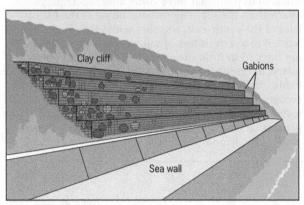

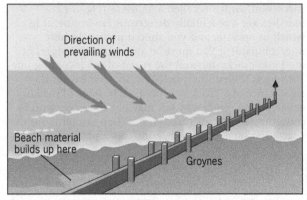

**Gabions** are wire mesh cages filled with boulders built on to the face of the cliff above a sea wall. They are only used where a settlement faces severe problems of erosion.

**Groynes** are built near sea walls and wherever longshore drift operates. They hold the beach in place and the beach protects the base of the cliff from erosion. Sand that has accumulated against one side of a groyne is usually redistributed over the beach by earth-moving machines.

**Costs: Revetments** £5000 per metre; **Sea walls** £10000 per metre; **Gabions** £1000 per metre; **Groynes** £20,000 each (placed 200 metres apart).

### Assessment of human impact

Two lines of approach are open to investigators of coastal protection: the first is the *assessment* of human impact, and the second is the *evaluation* of human impact (i.e. how successful the defences have been). There are a variety of methods for assessing human impact on coasts.

1   Archive material, including maps, Admiralty charts, photographs and postcards (see Section 3.8), may show the coast before protection was in place. A

comparison of protected and unprotected stretches of coast on present-day maps will reveal the effectiveness of the defence.

2   Survey and map the plan of the coast at the end of a sea wall. If the protection has been in position for some time, there may be a noticeable kink in the coast, indicating the approximate retreat of the non-protected section.

3   Reports by local authorities and nature conservation groups sometimes include research by consultants.

Any investigation must take into account the possibility that processes other than the obvious may be responsible for differential rates of coastal retreat. To give one example, at the same time as a promenade is built, groynes are erected on the beach, thus restricting longshore drift and the supply of material to the unprotected end of the beach. The amount by which the unprotected cliff has been eroded reflects not only the effectiveness of the promenade but also the accelerated erosion due to loss of sediment supply.

### Cost–benefit analysis

To evaluate the impact of coastal defence you need to use some form of cost–benefit analysis. Neither cost nor benefit is easy to quantify, and nor should you consider them purely in cash terms. Benefits include the following: protection of buildings, roads and railways, farmland (can this be justified at a time of farm surpluses and set-aside?), golf courses and nature reserves; preservation of sandy beaches, without which there would probably be loss of tourist earnings.

Costs include the financial cost of construction and maintenance, loss of visual amenity (few defences blend in: many are scars on the landscape, as Figure 6.13 illustrates), and (much harder to verify and evaluate) the long-term consequences, possibly to areas miles away if longshore supplies of coastal sediment are interrupted.

Any such analysis should aim to identify the key players or decision takers and assess the outcome from alternative perspectives. In the UK today there is sometimes a conflict of interest between the Environment Agency, which is in charge of protecting estuaries, and county councils, which are responsible for the rest of the coast. A council that installs walls and groynes to protect its own stretch of coast cuts off the supply of sediment to an estuarine salt marsh or spit.

Before the floods in England in early 2014 the Environment Agency recommended 'managed retreat', which means abandoning man-made sea defences and encouraging the expansion of natural defences such as salt marsh. This strategy also provides ecological habitat and is very much cheaper.

## 6.9 Sea-level change

Evidence of relative changes in the position of land and sea is preserved in some areas around the UK's coast. The emergent coastlines of western Britain, notably in Scotland and Wales, provide a splendid account of many of the changes of the last 10 000 years. It may be beyond your resources as a student to date raised beaches, but you should know what to look for and be able to find evidence to confirm the marine origin of a feature 25–30 m above the current sea level.

### Raised beaches

These can be found at a number of sites composed of resistant geology. Apart from the appearance of a raised beach (Fig. 6.14), evidence you should look for includes:

i a top covering of soil or peat;
ii possibly some head deposit: angular, frost-shattered debris;
iii rounded pebbles, including some 'erratics' of alien origin; mixed in with

iv sea shells, similar to those found on an active beach;
v wave-cut platform sloping down to the sea, although this may be masked by the overburden described above. Do not try to clear away the mantle of debris, but look for naturally exposed sections.

### Dead cliffs

Behind the raised beach you may find a line of old, inactive cliffs. A geographer is mostly interested in the extent to which the cliff face has declined since the cliffs were abandoned by the sea. The dominant process is no longer marine undercutting but sub-aerial denudation and mass wasting. Compare the form of the cliff profile with the active cliff below. Use a local guide or an authoritative text on coastal geomorphology to help you to date the abandonment of the cliff. Use this to assess (to the nearest 1000 years) the rate of decline of the old cliff.

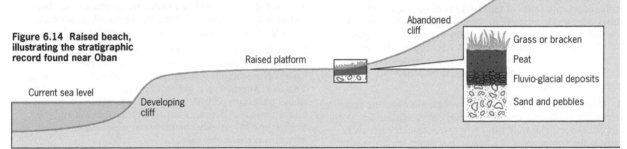

**Figure 6.14 Raised beach, illustrating the stratigraphic record found near Oban**

Current sea level

Developing cliff

Raised platform

Abandoned cliff

Grass or bracken

Peat

Fluvio-glacial deposits

Sand and pebbles

# 6.10 Project suggestions

Coasts are rich in project potential. The project you choose will depend on the nature of the coast most accessible to you. A number of outline project ideas are included in this chapter (above), while more detailed suggestions follow.

**1** Human impact on the coast takes many forms. Sometimes it is designed to achieve specific results (see Section 6.8 above) and sometimes not, as in the case of dumping mining spoil on beaches. You might decide to measure how effective sea defences have been. This can really only be done where a protected stretch of coast lies near an unprotected section. Use old OS maps, Admiralty charts or photographs (see Sections 3.7 and 3.8) to see what the coastline looked like before the sea defences were built, and compare with today's OS maps.

The profile of a cliff whose base is protected by a promenade will often look different from that of an unprotected cliff. You can examine cliff profiles at regular intervals along the coast to see what effect the sea defences have had. In all coastal protection projects, try to find out from the local authority when the sea defences were built. You can then calculate the speed at which subsequent changes have occurred.

Dumping of mining spoil on beaches has been carried out in parts of Britain for hundreds of years. Feeding the beach with large quantities of sediment has sometimes had a dramatic effect on the geomorphology and on the environmental quality of the area. In terms of the physical effect, coal mining waste has resulted in the rise of beach level and the protection of cliffs in much the same way that sea defences would have done. Compare the morphology of the cliffs affected by the mining spoil with the adjacent unaffected areas. It should be comparatively easy to find out in what year dumping began.

An alternative approach is to assess coastal defences from the cost–benefit perspective. Look at page 57 for suggestions on this type of project.

**2** Salt marshes have been disappearing in some places at an alarming rate. Although it is impossible to measure the rise of sea level responsible for this, it should be possible to trace the rate at which salt marsh is being eroded. Using old OS maps or Admiralty charts you should be able to measure by how much the marsh has retreated. Assuming the same rate of retreat, how long will it be before the marsh has disappeared? What will be the consequences of its disappearance? You should consider plant diversity and loss of habitat, or perhaps you can assess the importance of the marsh as a sea defence. What land uses lie behind the marsh, for example? If it is farmland or land used by tourists, what is the economic value of that land? Is there a sea wall or other man-made obstacle preventing the inland growth of the salt marsh? If so, are there plans to raise or strengthen it? What are the costs of doing so? Who is going to pay? What property is threatened by rising sea levels? What do the owners or occupants think should be done? What is the cost of rehousing these people?

**3** Coastal change is rapid, but whether you have the time necessary to observe change depends on individual circumstances. The minimum you can reasonably get away with is the comparison of a beach in summer with the same beach in winter. The aim of this project is to examine the effect of high-energy winter storm waves and low-energy summer conditions on the beach profile. Record wave energy, wave type, beach profiles, and particle size during the low-energy conditions of summer. Make these observations again at the same site following stormy winter weather and compare the two. How much more powerful were the storm waves? Do the profiles conform to the winter cut and summer fill pattern shown in Figure 6.4? Is this due to the presence of larger beach material in winter or to the action of different types of wave?

**4** Careful experimental design and site selection is necessary in projects involving cliffs. Two hypotheses might be *Cliffs that are actively undercut are steeper than those that have basal protection* and *Geological factors control cliff profile*. To test the first hypothesis, you could use a suitable stretch of indented coastline with exposed headlands and sheltered bay-head beaches, such as those along the south Cornwall or south Wales coast, but its geology must be uniform. To test the second hypothesis, you must find the reverse conditions, namely varied geology (structure and lithology – a concordant and discordant stretch, for preference) but with roughly the same degree of exposure to wave attack.

**5** Another beach project could attempt to identify the many influences on beach profile. The experiment needs to be designed to isolate key influences if possible. One hypothesis might be *Beach steepness is related to particle size* (Table 6.1). Particle size is sampled along transects perpendicular to the sea using the pebbleometer (page 27) or sieving (page 63) in the case of sand. Plot the data on a histogram and compare the median particle size of the different transects. The Mann–Whitney $U$ test will tell you whether the differences in stone size are statistically significant. Measure the long profile of the same transects. Measure the percolation rate at two or three points along the transects, and measure either phase differences or swash and backwash strength with the swingometer. Wave type, steepness and/or energy should also be measured at each transect, as they will also affect beach form.

**6** Examine the form of a spit with a view to testing a theory of spit formation. This will involve a systematic sampling frame to collect data on lateral sorting (evidence of long-term longshore drift). Reference to old maps (see Section 3.7) may reveal updrift erosion of the spit and downdrift deposition, which caused a rotation of the feature to become more swash-aligned and less drift-aligned. Does the spit have recurves? Are these the product of wave refraction or of waves arriving from a different direction? Long-term meteorological data will be needed to supplement fieldwork, and these can be obtained from the Meteorological Office (see page 165) or perhaps the local coastguard.

**7** Studies of sand dunes or salt marshes could test the hypotheses *Biodiversity increases away from the sea over time* or *Plant succession results from changing edaphic and environmental conditions*. Alternatively, you could compare a transect of soil and vegetation type with models such as Figures 6.10 and 6.12. Note any differences between your transect and the models, and investigate conditions in your area to explain why this might be the case.

# 7 Ecology and pollution

For clarity of presentation this chapter examines soils and vegetation separately. In practice, both soils and vegetation are very closely linked, and project work will often focus on the interrelationships between the two. A section on pollution follows.

## 7.1 Soils

Soil is composed of weathered rock particles, decayed plant matter, water and air. The nature of soil is strongly influenced over time by climate. Figure 7.1 summarises the factors that affect the development of soil.

This section explains some of the techniques used to investigate the composition, texture, and physical and chemical properties of soil. Although these soil characteristics are of interest in their own right, geographers often relate soil quality to patterns of vegetation and to their agricultural potential.

The British Geological Survey publishes soil maps and land cover/land use maps for the UK. County and district surveys (which include maps) are also available, and these may prove to be a useful source of secondary data.

### Soil sampling

As with any experiment requiring data collection, once the scientific objectives of the investigation have been decided then a sampling strategy must be drawn up. The sampling strategy is designed to ensure that the

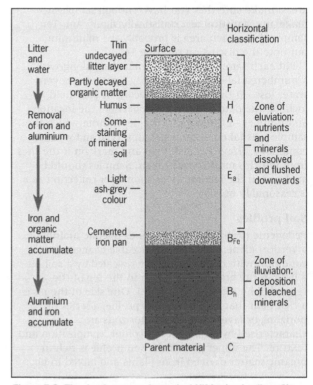

**Figure 7.2  The development of a typical UK lowland soil profile (podsol); this occurs where there is a net downward movement of water**

appropriate number of samples are collected from the appropriate locations in order that our question can be answered or our hypothesis tested.

Is there an environmental gradient in the area to be tested? In other words, is the area on a slope? Does distance increase from the sea? Is there an underlying steady change in geology? If there is, we collect samples along a transect running across the environmental gradient. If not – for instance, if we are comparing the soils of an orchard with those of a cornfield – we collect samples at random. These ideas are discussed in more detail in Section 7.3 and on pages 14 and 15.

The depth from which soil samples should be collected depends upon the scientific object. If you are investigating soil-forming processes, take samples from different depths in the profile. For investigations of soil quality along a transect, collect your samples from a consistent depth in the L–E horizons (Fig. 7.2).

The number of soil samples collected must reflect your scientific object. If a transect survey is being conducted, then samples should be collected at regular

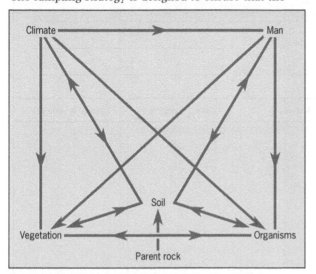

**Figure 7.1  Interrelationships between different pedogenic processes**

intervals. Note, however, that experimental error and the relatively inexact laboratory tests available to you mean that subtle variations in soil quality will simply not be measurable. Therefore, do not take samples too close together unless you are sure of a sharp variation in soil quality. If comparing the soils of two different areas, make sure that you have enough samples to make any statistical test statistically significant. Ten samples from each area is probably the minimum, but note the discussion on page 16.

Put each sample into a polythene bag, along with a numbered label, and tie the bag tightly to prevent water loss. At the same time, record in your field logbook the date and time of collection, the location and number of the sample, the depth from which the sample was taken, local vegetation type and weather conditions. Draw a sketch map and mark on it the sites from which samples were taken. Samples should be analysed in the laboratory as soon after collection as is reasonably possible.

### Soil profiles

Pedogenic (soil-forming) processes operate mainly in a vertical plane, producing noticeable changes in soil condition with depth. These are revealed by a soil pit, a hole in the ground dug down to the top of the regolith (the weathered mantle). One side of the pit is a clean cut that reveals the soil profile, a series of horizons or layers in the soil. Horizons are characterised by differences in colour, composition and texture. The upper part of the soil profile is rich in organic matter derived from plants and mixed in by

earthworms. The lower soil horizons are more stony, because of the proximity of the bedrock. In an undisturbed soil (i.e. not ploughed), you should be able to identify several distinct horizons. These layers are, by convention, given letters to help with the comparison of soil types. Figure 7.2 illustrates and explains a typical UK soil profile.

*Soil pits*
1 Permission must be obtained from the landowner before you begin digging.
2 If the area is vegetated, cut the turf and place it to one side.
3 Place the dug soil on a plastic sheet.
4 A pit is normally 1m deep, or down to the regolith (the weathered mantle), whichever is the less.
5 Slice one face clean.
6 Measure the thickness of each horizon in your field notebook or on a recording sheet, such as that shown in Figure 7.3.
7 Fill the pit and restore the vegetation.

## Soil composition

### Soil water content

Soil water is important in many ways: it provides plants with moisture and acts as a medium of transport for nutrients. Soil moisture is clearly essential for plant growth. However, waterlogged soils are poor, as water occupies nearly all the pore space and thus excludes air from the ground. Many micro-organisms that break down organic matter are aerobic and need gaseous

### Data sheet

Profile location _____     Date _____

Profile number _____     Time _____

Weather conditions _____     Vegetation cover _____

| Sample depth (cm) | pH | Bulk density | Texture | Loss on drying (%) | Loss on ignition (%) |
|---|---|---|---|---|---|
| 1 | | | | | |
| 2 | | | | | |
| 3 | | | | | |
| 4 | | | | | |
| 5 | | | | | |

Figure 7.3 Sheet for recording field and laboratory data in soil analysis

oxygen. They cannot survive in anaerobic conditions, so the recycling of plant matter is reduced. The method for measuring soil water content is as follows.

1  Crumble a small amount of soil (around 50 g) into a small pre-weighed crucible and weigh the crucible plus soil.
2  Put the sample in an oven at 105 °C overnight to remove the water.
3  Take the sample from the oven. Let the crucible cool down. Weigh the crucible and dry soil together.
4  Subtract the weight of the crucible to give the dry soil weight.
5  Subtract the weight of the dry soil from the weight of the wet soil to give the weight loss on drying (total weight of water evaporated from the soil).
6  Substitute your figures in the following equation:

$$\frac{\text{weight loss on drying}}{\text{weight of wet soil}} \times 100 = \begin{array}{l}\text{percentage water} \\ \text{(by weight)} \\ \text{contained in the soil}\end{array}$$

Soil water content will vary from day to day with weather changes, from place to place (for example, down a slope), from soil type to soil type, and with position in the soil profile.

## Soil organic content

When plants die and are left to rot, bacteria rapidly cause their decay. The nutrients that the dead plants once took from the soil are returned to the earth. If the soil is to remain fertile, then this organic matter has to be recycled. As we shall shortly see, however, high concentrations of organic matter do not necessarily mean highly fertile soils.

To determine the weight of organic matter in a soil sample:

1  Take 5–10 g of dry soil from the previous experiment and grind it with a pestle and mortar.
2  Place the ground-up soil in a pre-weighed crucible and weigh the soil and crucible.
3  Burn the soil sample in the crucible over a bunsen burner for half an hour. This should ensure the incineration of all organic matter.
4  Let the crucible cool down. Wipe off any carbon deposited on the underside of the crucible. Weigh the burned soil plus crucible and subtract the known weight of the crucible.
5  Subtract the weight of the soil after burning from the weight of the soil before burning to give the weight loss on burning (i.e. the weight of organic matter).
6  Substitute your figures in the following equation:

$$\frac{\text{weight loss on burning}}{\text{weight of dry soil}} \times 100 = \begin{array}{l}\text{percentage organic} \\ \text{matter (by weight)} \\ \text{of dry soil}\end{array}$$

The concentration of organic matter in soil varies considerably. On arable land, topsoil values lie between 1 per cent and 5 per cent, whereas peats, often of low fertility, contain up to 80 per cent organic matter. The volume depends on the balance between the rate of gain and the rate of loss of matter. Certain vegetation, for instance deciduous woodland, supplies large amounts of organic matter, whereas root crops supply very little. Loss of organic matter depends largely on the rate at which it is broken down. Waterlogged soils with low oxygen content contain few organisms; the breakdown of dead vegetation is therefore slow and the concentration of organic matter may become quite high – such as peat. Well-drained and aerated soils allow a more rapid breakdown of dead vegetation and a more speedy recycling of nutrients.

## Soil chemistry

The chemical composition of soil is determined by the type of parent rock, soil texture, vegetation type, rainfall totals and average temperatures, and in many places by centuries of agricultural activity. The single most important chemical characteristic of soil is its acidity (as indicated by pH). The pH value of soil is a measure of the concentration of hydrogen ions in the soil water. The greater the concentration of hydrogen ions, the more acidic soil is. In most soils pH values lie between 5 and 9; pH 7 is neutral, lower values are acidic and higher values alkaline.

Acidic soils are often found in areas of high rainfall. The downward movement of water through the soil leaches or flushes out bases, leaving a high concentration of hydrogen ions. Acidic parent rock, notably igneous and metamorphic rock, also produces acid soil. Alkaline soils are found in areas of calcareous geology – limestone and chalk. However, leaching and vegetation can reduce the natural alkalinity and even produce weakly acidic topsoils in limestone areas.

Soil texture also influences pH. Bases are leached rapidly from sands, whereas clays with their smaller surface area and powerful electrical bonding retain them much more effectively. Sandy soils tend, therefore, to be more acidic than clays.

The pH value of soil does not directly affect plant growth except in cases of extreme acidity. It is important, however, because it affects the availability of nutrients for plant growth. Some nutrients essential for crop health (e.g. iron and magnesium) become scarce at pH values of 7.5 and over. At the opposite end of the scale, in acid soils below pH 5.0, the concentration of minerals such as aluminium and iron, mobilised by the acidity, may prove toxic. Some plant species prefer acidic soils, and others neutral or chalky soils. Among the former are gorse, bracken, brambles, rhododendrons, heather and pine trees. Some species, particularly gorse and heather, actually promote soil acidity through ion exchange.

DETERMINATION OF pH: COLORIMETRIC METHOD Of the techniques available for the determination of soil pH, perhaps the most robust and reliable is the colorimetric method. A BDH barium sulphate soil test kit can be used or, alternatively, buy a pH test kit from a garden centre. To determine pH using the BDH kit:

1 Pour a 5–10 g sample of soil into a test tube.
2 Add barium sulphate, distilled water and universal indicator.
3 Put in the bung and shake the tube well. Leave to stand until a clear, coloured fluid develops.
4 Compare the colour of the fluid with the chart provided and read off the pH to the nearest 0.5 of a unit.

Although interpreting colours may not be as precise as reading the dial of an electronic pH reader, such instruments are not always accurate. The colorimetric method allows samples to be read in the field, it is quick, cheap, reliable and easy to use.

pH METERS Electronic instruments for use in the laboratory or in the field are readily available. Their advantage is precision, to the nearest 0.1 of a pH unit. However, natural fluctuations greater than this may occur as a result of soil moisture content and temperature changes, so small variations in recorded pH must be treated with caution.

Read the manual accompanying your meter to find out exactly how to use it. The most important rule is to recalibrate the meter by use of buffer solutions of known pH between each reading. Do not contaminate the buffer solution: if you suspect it has been affected by the solution you are measuring, then change it.

NITRATES Nitrates as a pollutant are dealt with on page 77. Nitrogen from the atmosphere is fixed as nitrate in soil. Nitrate is also added as a chemical fertiliser by farmers. Soil nitrates are measured with nitrate test strips, but the nitrate must first be leached from the soil.

1 Place 50 g of soil in a beaker and top up to the 100 ml level with deionised water.
2 Stir the soil and water together and pour the mixture through a filter paper placed inside a funnel.
3 Test the filtrate with a test strip and refer to the colour chart to find the concentration of nitrate in mg/l, or parts per million (ppm).
4 Take a sub-sample of the soil and measure the percentage soil moisture (page 60); nitrate concentration can now be expressed as milligrams per gram of dry soil.

Nitrate concentrations vary seasonally and spatially. The concentration of nitrate in soil is, of course, high after nitrate fertilisers have been added, and lower after heavy rain. However, rain and soil moisture flush nitrates downhill. The concentration can therefore become high at the foot of hills. Nitrates are often leached into groundwater or directly into ponds and streams, where eutrophication occurs (page 77).

## Soil texture

Soil texture refers to the size of particles that make up the soil. There are three categories of soil defined by texture: sands, loams and clays. Sandy soils are composed mainly of fairly coarse grains (less than 2 mm diameter). They are low in nutrients but are light, well drained and well aerated. Clay soils consist largely of particles less than 2 $\mu$m ($2 \times 10^{-3}$mm) diameter. They are rich in nutrients, sticky, and often waterlogged. Loamy soils (see Fig. 7.4) are a combination of sand ($\geq$40%) and clay ($\geq$20%). Loams make excellent agricultural soils, being both well drained and fertile.

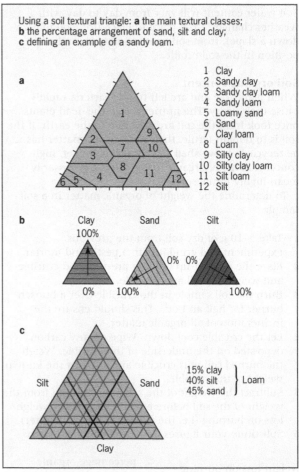

**Figure 7.4 The soil textural triangle: (a) the main texture classes; (b) directions for reading sand, silt and clay; (c) defining a loam**

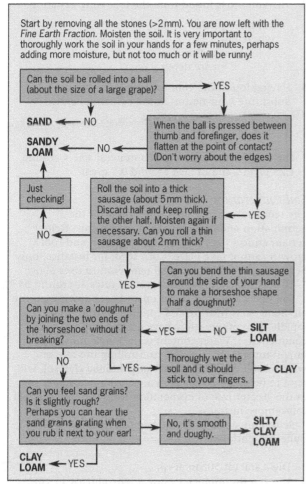

Start by removing all the stones (>2 mm). You are now left with the *Fine Earth Fraction*. Moisten the soil. It is very important to thoroughly work the soil in your hands for a few minutes, perhaps adding more moisture, but not too much or it will be runny!

Can the soil be rolled into a ball (about the size of a large grape)? → YES

SAND ← NO

SANDY LOAM ← NO ← When the ball is pressed between thumb and forefinger, does it flatten at the point of contact? (Don't worry about the edges)

Just checking!

NO

Roll the soil into a thick sausage (about 5 mm thick). Discard half and keep rolling the other half. Moisten again if necessary. Can you roll a thin sausage about 2 mm thick? ← YES

YES → Can you bend the thin sausage around the side of your hand to make a horseshoe shape (half a doughnut)?

Can you make a 'doughnut' by joining the two ends of the 'horseshoe' without it breaking? ← YES ── NO → SILT LOAM

NO

└ YES → Thoroughly wet the soil and it should stick to your fingers. → CLAY

Can you feel sand grains? Is it slightly rough? Perhaps you can hear the sand grains grating when you rub it next to your ear! → No, it's smooth and doughy. → SILTY CLAY LOAM

CLAY LOAM ← YES ─┘

**Figure 7.5  The determination of soil texture by hand**

### Hand texturing

The simplest and quickest way of assessing soil texture is to model a wetted sample in your hand. Begin by removing all stones of more than 2 mm, moisten the soil and mould it in your hand for a few minutes, then follow the steps shown in Figure 7.5.

### Sieving

A more sophisticated method of assessing soil texture is sieving, but this obviously depends upon the availability of a soil sieve.

1  Take the dried sample from the soil moisture experiment and carefully break any lumps in a mortar with a rubber-tipped pestle. (Alternatively, the soil can be soaked in water and Calgon to disperse the lumps.)
2  Place the sample in the coarsest sieve at the top of the nest of sieves and put the lid on.
3  Sieve for 10 minutes (Fig. 7.6).

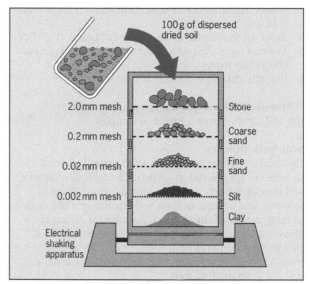

100 g of dispersed dried soil

2.0 mm mesh — Stone
0.2 mm mesh — Coarse sand
0.02 mm mesh — Fine sand
0.002 mm mesh — Silt
— Clay

Electrical shaking apparatus

**Figure 7.6  Measuring soil texture by sieving**

4  Empty the contents of each sieve into a weighing boat. Remove the finest particles with a brush, and weigh.
5  Plot a histogram of weight against particle size (phi scale – see page 27), or a cumulative percentage graph on probability paper. The advantage of this last method is that a number of samples can be plotted on the same axes for ease of comparison.

A major advantage of sieving is that it gives you quantitative data. Drawbacks are: it is inaccurate at the clay-sized end of the spectrum; it should only be used with soils of less than 5 per cent organic content.

### Sedimentation

A method that avoids the two drawbacks of sieving is sedimentation. Its greatest appeal is its technical simplicity. The method is based on the premise of Stokes' Law (the velocity of a particle falling through a viscous medium is directly proportional to the diameter of the particle). The settling speeds of the different fractions allow you to compare them visually.

1  Add 50–100 g of separated soil to a measuring cylinder and fill it with water.
2  Place your hand over the end and shake.

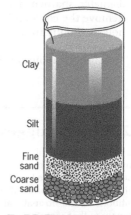

Clay

Silt

Fine sand

Coarse sand

**Fig 7.7  The determination of soil texture by sedimentation**

**3** Allow the fractions to settle undisturbed for at least 8 hours.

**4** Read off the depths of the different bands (Fig. 7.7).

**5** Express each class as a percentage of the depth of the soil in the cylinder.

**6** Plot the data as a divided histogram.

The main difficulty is applying a 'constant eye' to the soil bands across a range of samples.

**Soil bulk density**

Bulk density is the weight of a soil sample divided by its volume (expressed in g/cm³). The bulk density of a soil sample reflects three things: the proportions of mineral and organic matter, soil texture and the degree of compaction of the ground. As mineral matter is dense (2.5 g/cm³) compared with organic matter (0.5 g/cm³), mineral-dominated soils with a high proportion of weathered rock particles have a higher density than organic soils.

Bulk density is strongly affected by land use and soil management. Ploughing and tilling help to break up the soil and increase the air content, but the use of heavy tractors and other modern farm machinery such as disc ploughs compacts the ground and damages the soil crumb structure. Traditional organic fertilisers (farmyard manure) improve crumb structure by helping to bind the soil grains together, which chemical fertilisers cannot do. Soil fertility depends upon a well-aerated soil supporting a dense population of soil organisms such as bacteria and earthworms.

Bulk density increases with depth. This reflects decreasing proportions of organic matter, less root penetration and compaction by the overlying soil. The method for measuring bulk density is as follows.

**1** Take a soil corer or auger: this is a metal cylinder, open at one end, which you screw about 6 cm into the soil. If you cannot get a soil auger, use a short length of sturdy plastic piping, about 3–4 cm in diameter, sharpened at one end.

**2** Remove the corer and measure the depth of the hole in centimetres. The core itself will have been compressed.

**3** Measure the radius of the soil core in centimetres.

**4** The volume is found from the following equation:

$$\text{volume} = \pi r^2 d$$
where $\pi = 3.14$
$r$ = radius
$d$ = core depth

**5** Having found the volume, weigh the sample. Substitute your figures in the equation below:

$$\text{bulk density} = \frac{\text{weight}}{\text{volume}}$$

The average bulk density of a mineral soil is around 1.25 g/cm³, and of peat around 0.5 g/cm³.

*Soil temperature*

The temperature of a soil is important for the germination and growth of seeds and plants. Below certain critical temperatures, germination and root growth cannot take place. Corn seed, for instance, only begins to germinate once soil temperature rises above 7–10 °C. It reaches optimum growth rates at around 35 °C.

Soil takes longer to warm up and cool down than the air. Some soils heat up and conduct warmth more rapidly than others (they have greater thermal conductivity). Water content is the single most important soil characteristic controlling the rate at which a soil will heat up. A badly drained clay soil will tend to be much cooler than a well-drained sandy soil, as the greater rate of evaporation from the clay soil will cause more heat loss.

Setting up a site to investigate soil temperature is quite straightforward.

**1** Dig a soil pit 50 cm deep.

**2** Insert a thermometer horizontally at least 15 cm into one side of the pit at 2 cm depth. Add two further thermometers at depths of 10 cm and 30 cm.

**3** Cover the pit with a lid to keep out the sun's rays.

**4** Take temperature readings at regular intervals over a 24-hour period – as late as possible at night, and as early as possible before sunrise. Record the air temperature at the same time that you record the soil temperature.

**5** Plot the readings of the three thermometers on one graph. Observe the lag behind the air temperature that each soil depth displays.

# 7.2 Project suggestions: soils

You must think carefully about experimental design and the sampling frame. You need to be careful for two reasons. Firstly, most soils have been altered dramatically by farmers, so that little of their original condition remains. Secondly, soil conditions do not normally alter perceptibly over a short distance, so that sampling and experimental error are often greater than natural variation.

**1** Spatial variations in nitrate level can be investigated quite easily. The aim is to determine which patterns of land use and topography produce the highest concentrations of nitrates. Choose a study area within a single drainage basin, and draw up a sampling frame to decide the number and location of sample sites. You need to measure soil nitrate under different land uses, such as woodland, moorland, pasture, cereals and

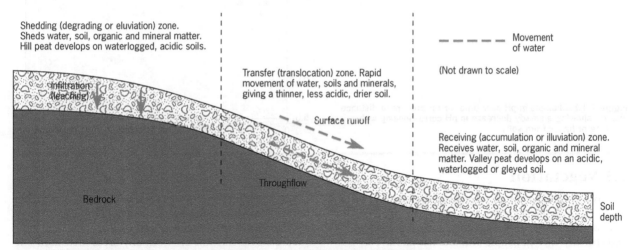

Shedding (degrading or eluviation) zone.
Sheds water, soil, organic and mineral matter.
Hill peat develops on waterlogged, acidic soils.

Infiltration (leaching)

Transfer (translocation) zone. Rapid
movement of water, soils and minerals,
giving a thinner, less acidic, drier soil.

Surface runoff

– – – – – Movement
of water

(Not drawn to scale)

Receiving (accumulation or illuviation) zone.
Receives water, soil, organic and mineral
matter. Valley peat develops on an acidic,
waterlogged or gleyed soil.

Bedrock

Throughflow

Soil
depth

**Figure 7.8 A soil catena, showing the relationship between soil, slope angle and hydrology**

root crops (stratified random or stratified systematic). Use a systematic sampling frame to determine measuring sites at different gradients and across the catchment.

Take several readings at each site. Map the results. You will probably find a close connection between the concentration of nitrates and the intensity of farming. If the farmer has let you sample on his land, he may also be willing to let you know the doses of nitrate fertiliser he has added to different fields. However, you will probably observe that patterns of nitrate are more complex than this. In fact, the movement

of soil moisture may have led to nitrate leaching from some areas and concentration of nitrates elsewhere. What do you learn from looking at infiltration rates and soil texture? Map the surface hydrology of the study area. What light does this throw on the pattern of nitrates? What are the hydrological properties of the bedrock?

**2** Quite dramatic changes in soil quality can be seen down a hillside. Soil change down a slope is called a catena. Soil-forming processes operate at different rates at positions down a slope; Figure 7.8 summarises some of the observable changes that you might look for. Indeed, this diagram could serve as a model for you to compare with your own findings. You should observe changes in soil quality down the hillside and also try to measure the environmental changes (climatic and hydrological) that lead to variations in the soil-forming process.

**3** You can examine the influence of microclimates on soil formation on opposing sides of a valley running east–west. The warmer south-facing slope would be expected to have deeper soils (although this will also be influenced by slope angle). Weathering processes normally operate more effectively on a south-facing slope and produce a finer soil texture. Measure soil depth and collect samples at regular intervals along a transect on both sides of the valley. Measure the slope angle at each site (Fig. 7.9). In order to help confirm your hypothesis, record temperature (air and soil) on both sides of the valley (see page 81). Both climatic and soil data need to be tested statistically, perhaps using the Mann–Whitney $U$ test (page 147) to see whether any apparent differences are statistically significant.

**4** With reference to sand dunes (Section 6.6), test the hypothesis: *Soil acidity increases over time.* Samples need to be taken at regular intervals along a transect from the beach going inland. While it may not be possible to date the dunes with any exactness, you may get some idea of dune age from old maps (Section 3.7), local publications or (in the case of dunes in a nature reserve) from the Park Warden. Use Figure 7.10 as a model with which to compare your results. Leaching and the role of plants in the process of acidification should be considered as explanatory factors.

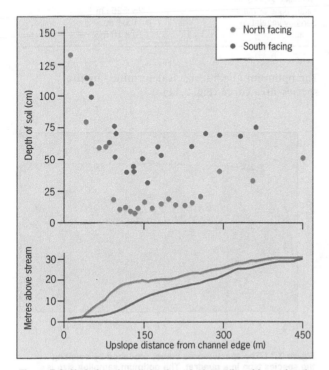

**Figure 7.9 Variations in soil depth on facing valley sides, related to aspect and slope angle**

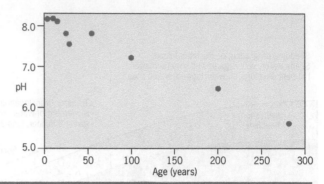

**Figure 7.10 Changes in pH over time (or for time, read distance inland), showing a steady decrease in pH corresponding with leaching of bases from soils**

## 7.3 Vegetation

The interest of vegetation to most geographers lies in the way plants adapt and respond to their environment. Microclimate, altitude, aspect, soil type and the activities of people and animals all have a noticeable influence over the type and condition of vegetation. The techniques described in this section are chiefly concerned with the sampling and analysis of plant communities, with a view to relating them to the above factors.

### Vegetation sampling

Sampling must be appropriate to the nature of the vegetation and the environmental influences acting upon it. Where there is an environmental gradient – a steady change in soil, microclimate or altitude, for example – we use systematic sampling, normally along a transect. This would apply to studies of changes through a wood, across sand dunes or salt marshes or even across a footpath. The transect is a line along which samples are collected at regular distances or altitudes.

If the distance is short – 30 m across a single dune, for instance – sampling can be continuous. If the distance is greater, samples are collected at intervals appropriate to the distance, the rate of change in vegetation type and the time available.

Where there is no discernible environmental gradient, random sampling is appropriate and would be used to compare two areas of contrasting vegetation, on chalk and clay, for example. If you are investigating the vegetation of an area that is not uniform, because some lies on chalk and some on clay or because one stand of woodland is much older than the rest, it is necessary to stratify the sampling with the number of samples taken from each area in accordance with its size. Stratified sampling should also be used if species grow in clumps or clusters, as they may be underestimated or omitted altogether if using a purely random method. See Section 2.3 for further discussion of sampling.

### Quadrats

The size and type of quadrat depends upon the type of vegetation being sampled, as illustrated in Table 7.1.

Usually, quadrats of 1 m² and below are wire or wooden frames, sometimes with string or metal crosswires every 10 cm to aid in the process of estimating area coverage. Quadrats over this size are normally areas measured out on the ground by tape.

A point quadrat is a frame with holes at regular intervals (often 10 cm apart) through which a long pin is passed; this method is used in places where the vegetation overlaps. Every species touched by the pin as it goes down to the ground is recorded.

**Table 7.1 Suggested quadrat sizes**

| Type of vegetation | Size of quadrat |
| --- | --- |
| Mosses and lichens | 10 x 10 cm |
| Woodland ground flora, heath, dune or marsh | 25 x 25 cm |
| Shrubby vegetation | 1 x 1 m |
| Trees | 10 x 10 m |

The optimum quadrat size is determined by the species–area curve (Fig. 7.11).

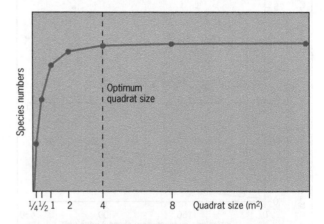

**Figure 7.11 The species–area curve, produced by counting the species type in a quadrat. The optimum sampling size is indicated; this is where a further doubling of quadrat size has an insignificant effect on the number of new species recorded.**

*Species–area curve*
1  Count and record the number of plant species within a 0.25 m² quadrat.
2  Double the quadrat area and count the number of new species. Add this number to the earlier total.
3  Repeat the process until no new species are found.
4  Plot the data.
5  The optimum quadrat size is that beyond which you judge any more species to be unimportant, as shown in Figure 7.11.

The relative spatial importance of different plant types can be measured in two ways, by cover and by frequency. Cover means you have to estimate the proportion of a quadrat occupied by a given plant type (difficult in large quadrats), while for frequency you must calculate the percentage of quadrats in which each species is present (regardless of the area covered or the number of individual plants). The second method tends to underestimate species that grow in clumps.

For statistical purposes, the siting of the quadrat should be random, but such an approach may result in large areas being left unsampled. This problem can, to some extent, be overcome by systematic random sampling: divide the study area into large grid squares and sample at random in every square. Whichever method you use, ensure that your sample is representative of the background population, if necessary by using the species–area curve.

Note that a number of quadrat samples can be substituted for the size of a quadrat – in other words, instead of making the quadrat bigger, simply take more small quadrat samples until the running total of new species encountered begins to tail off.

**Transects**
Quadrats may prove difficult for sampling larger plant types such as trees, so sample along a transect which runs perpendicular to your main transect at regular intervals. The length of this transect will depend upon the size and spacing of the trees, but you could easily produce a species–length curve to calculate the optimum length. Every 1 or 2 metres, note the species of tree closest to the metre mark. Count a tree only once if it falls exactly between two adjacent metre markings.

*Identification*
Identification of the flora of the British countryside often requires a guide book, particularly in winter when plants have lost their leaves and flowers. There are a number of plant guides available. Figure 7.13 is a simple field key to the more common hedgerow shrubs.

*Recording plant types*
The relative spatial importance of different plant types can be measured in two ways, by cover and by frequency. Cover means that you have to estimate the proportion of a quadrat occupied by a given plant type (difficult in large quadrats). It is common for the total to add up to more than 100 per cent as plants overgrow each other. Frequency counts can be made by calculating the percentage of quadrats in which each species is present (regardless of the area covered or the number of individual plants). This method can be improved upon by dividing the quadrat into smaller squares with string, recording the number of times each species is present in each small square, and calculating this as a percentage of the total number of small squares in all the quadrats sampled. Percentage frequency results can be plotted as kite diagrams, as Figure 7.12 shows, for ease of analysis.

Occasionally, we need to know the number of individual plants present, in which case use the point quadrat described above. Count the number of plants of each species that touch the pin. This allows analysis of diversity, using Simpson's Diversity Index, for example.

**Figure 7.12  Kite diagrams showing percentage frequency of species up the Gann salt marsh (*Source:* K Black, *Geography Review*, September 1998)**

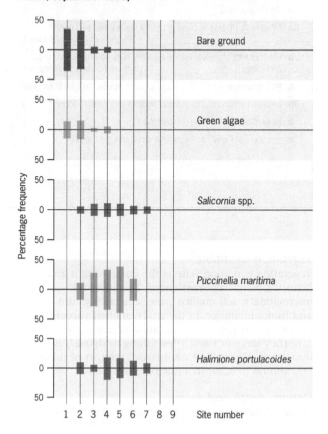

Figure 7.13 A simple field key to the more common hedgerow shrubs

# A simple field key to the more common hedgerow shrubs

**la** Shrub evergreen 2

**b** Shrub deciduous 3

**2a** Leaves prickly Holly (*Ilex aquifolium*)

**b** Plant climbing or twining; leaves not prickly Ivy (*Hedera helix*)

**c** Not as above (Collect for identification)

**3a** Shrub with weak climbing or trailing stems 4

**b** Shrub erect or semi-erect and straight 7

**4a** Stems armed with numerous prickles
Blackberry (*Rubus fruticosus*)

**b** Stems unarmed 5

**5a** Shrub with rope-like stems, climbing by twisted leaf stalks; often with the remains of last year's hairy fruit
Old Man's Beard (*Clematis*)

**b** Shrub with twining stems, thin and not fibrous 6

**6a** Upper leaves with two small lobes at their base; flowers purple, small Woody Nightshade (*Solanum dulcanara*)

**b** Upper leaves entire, without stalks Honeysuckle (*Lonicera*)

**7a** Shrub semi-erect, armed with prickles at nodes
Gooseberry (*Ribes uva-crispa*)

**b** Shrub erect with single stout spines at branch ends and nodes 8

**c** Shrub unarmed 9

**8a** Flowers April–May appearing before leaves; leaves entire
Blackthorn (*Prunus spinosa*)

**b** Flowers late May–July appearing after leaves; leaves deeply cut and lobed Hawthorn (*Crataegus mongyna*)

**c** Not as above (Collect for identification)

**9a** Leaves compound, made up of leaflets in a pinnate form 10

**b** Leaves simple with ± irregular lobes 11

**c** Simple, not lobed; entire or with small marginal teeth 12

**10a** Buds large, black; twigs grey with smooth bark; flowers in purple clusters appearing before leaves; leaves finely toothed
Ash (*Fraxinus*)

**b** Buds smaller, but not black; twigs pale brown with rough scaly bark; flowers in white umbels appearing after leaves; leaves coarsely toothed Elder (*Sambucus*)

**11a** Leaves with small irregular rounded lobes; twigs glabrous often infested with round marble galls Oak (*Quercus* spp.)

**11b** Leaves uniform green on both sides with 3–5 acute irregularly toothed lobes; twigs greyish, glabrous
Guelder Rose (*Viburnum opulus*)

**c** Leaves paler beneath with 3–5 ± regular, entire lobes; cordate at the base; twigs brown and hairy Field Maple (*Acer campestre*)

**12a** Twigs bearing erect or pendulous catkins; leaves with small leafy projections (stipules) at base of leaf stalk 13

**b** Not as above 14

**13a** Leaf buds with numerous scales; twigs clothed with reddish glandular hairs; leaves 5–12 cm, suborbicular, toothed, cordate at base Hazel (*Corylus*)

**b** Leaf buds with single scale; twigs not as above; leaves various, not suborbicular or cordate at base Willows (*Salix* spp.)

**14a** Leaves opposite, entire, not toothed, with few indistinct veins 15

**b** Leaves alternate, coarsely toothed; veins and cross-veins outstanding 16

**15a** Twigs green, 4-angled; flowers not in umbels
Spindle Tree (*Euonymus*)

**b** Twigs purplish red – at least on the sunny side; flowers in flat-topped umbels Dogwood (*Thelycranea sanguinea*)

**16a** Leaves symmetrical at base, oval or ovate, densely hairy beneath; flowers in creamy-white umbels
Wayfaring Tree (*Viburnum lantana*)

**b** Leaves asymmetrical at base, hairy only in axils of main veins on the underside; flowers appearing before leaves, purple
Elms (*Ulmus* spp.)

**c** Not as above (Collect for identification)

## Vegetation condition

Vegetation condition reflects the environment and history a plant has experienced. This includes microclimate, soil quality, water availability, and animal and human damage. In the study of plant succession (seres), for instance, the condition of marker species (are they large or small? flourishing or dying?) is important. Figure 7.14 summarises the sort of things you should look for in a microclimate study.

## Canopy depth and density

Some studies require a measurement of the density of a wood, in order to relate it to factors such as microclimate. For example, the drag effect of trees on wind is related to tree height and the density of the canopy. Tree height can either be estimated, or determined by using trigonometry (page 52). The top and bottom of the tree canopy and the height of the other vegetation layers – the shrub layer, the field layer (grasses) and the ground layer (mosses, etc.) – can be measured and plotted diagrammatically (Fig. 7.15). The density of the tree canopy can be estimated visually or measured by using a light meter.

## Data sheet

Date _____     Slope aspect _____

Transect number _____     Slope angle (degrees) _____

| Flower/tree species | Height | Girth (chest level) | Leaf condition (estimated %) | | | | | | Flower condition (estimated %) | | | | | |
|---|---|---|---|---|---|---|---|---|---|---|---|---|---|---|
| | | | bud | bursting | just out | full out | falling | bare | bud | bursting | just out | full out | falling | bare |
| | | | | | | | | | | | | | | |
| | | | | | | | | | | | | | | |
| | | | | | | | | | | | | | | |
| | | | | | | | | | | | | | | |

**Figure 7.14  Data sheet for recording the condition of plants. Plant conditions recorded here will vary along a transect, reflecting either microclimate or other environmental factors such as soil quality.**

**Figure 7.15  The structure of an English woodland, illustrating how the density of the canopy and other layers can be shown**

Tree layer (canopy)

Shrub layer

Field layer

Ground layer

## Analysis of plant diversity

The ecological colonisation of an environment such as a freshwater lake or a sand dune follows a number of seral stages. Each sere is dominated by a particular vegetation community. Table 7.2 describes the stages of ecological succession. In the early stages plant diversity (the number of species present) increases. In the later ecological stages diversity tends to decrease as certain species predominate. There are a variety of ways in which diversity, or its opposite – dominance – can be measured, but one of the more useful and straightforward to calculate is Simpson's Diversity Index. The index lies in the range 0–1 where 0 is minimum diversity and 1 is maximum. For

comparative purposes, sample sizes should be similar.

$$d = 1 - \frac{\Sigma(n(n-1))}{N(N-1)}$$

Where:
$d$ is Simpson's Diversity Index
$\Sigma$ is the sum of
$n$ is the number of individuals of each species present
$N$ is the total number of samples taken

Let us calculate the diversity of Area A in which we have sampled the following numbers of plants:

| Species | Area A |
|---------|--------|
| A | 50 |
| B | 25 |
| C | 12 |
| D | 6 |
| E | 3 |
| Total | 96 |

$$\frac{50(49) + 25(24) + 12(11) + 6(5) + 3(2)}{96(95)} = 0.35$$

$\therefore$ Simpson's Diversity Index = $1 - 0.35 = 0.65$

**Table 7.2  Stages in ecological succession**

**Stage 1 Migration**  Seeds are carried by wind, water, animals or man to the site.
**Stage 2 Colonisation**  Pioneer species (often highly adapted) colonise in what is normally a hostile site.
**Stage 3 Establishment**  New species arrive and grow in the new soil. Conditions ease.
**Stage 4 Competition**  Equilibrium species replace pioneer species as competition increases for light, space and water.
**Stage 5 Stabilisation**  Competition diminishes as few new species arrive. Decline in number of species present.
**Stage 6 Climax**  The dominant vegetation community is established and in equilibrium with the environment.

# 7.4 Project suggestions: vegetation

It is difficult to isolate any single influence on vegetation. Soil quality, geology, hydrology, slope angle and human influence all play a part, but unlike laboratory scientists geographers cannot conduct controlled experiments. In testing a single hypothesis, therefore, you must be aware of the extent to which factors other than the one under investigation play a part. Where possible, however, you should design controlled experiments to isolate other influences.

**1**  The influence of microclimate on vegetation is particularly strong in early spring. An east–west-oriented valley provides an ideal opportunity for studying phenology (the influence of climate on vegetation). Observations of soil temperature, air temperature, radiation and wind speed should be made on the north- and south-facing slopes. Take these readings early in the morning, at lunch-time and late in the afternoon over a period of four or five days. Take one or more transects on each side of the valley and record plant type and condition (Fig. 7.14). The microclimates of the two sides may be different enough to produce differences in the condition of vegetation (for instance, in the number of trees in bud or full leaf). You might use a test of significance (page 147) to see whether the difference in species composition or in the number of trees in bud is truly significant or a matter of chance. The results can then be related to the climatic data.

**2**  In Chapter 6 we suggested studying plant successions (seres) in sand dunes and salt marshes. Depending upon where you live, you could study two other environments: plant successions on bare rock (lithosere) and in fresh water (hydrosere).

Limestone pavements offer a wide diversity of habitat, with most vegetation growing in the sheltered conditions of the grykes. The composition of the vegetation here will depend upon the width and depth of the gryke, and hence upon its microclimate, and the access of grazing animals such as

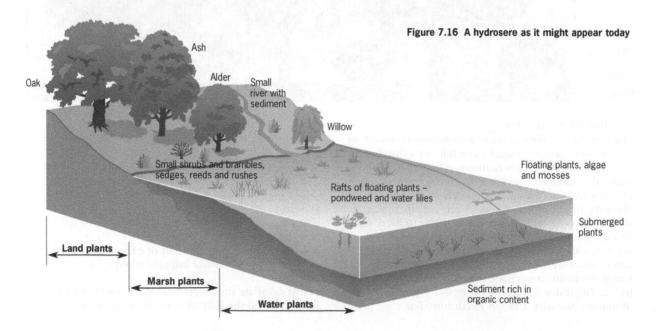

**Figure 7.16  A hydrosere as it might appear today**

rabbits. The lithosere will run from the edge of the pavement (dominated by grazed grassland) in towards the centre. Compare seres in pavements of different sizes.

Primary hydroseres are found in newly formed freshwater ponds: the meres of Cheshire and other kettle holes are suitable (but not abandoned gravel pits and flooded quarries, which are deep, steep-sided and very dangerous). Figure 7.17 shows a typical sequential pattern, ending in the almost total sedimentation of the pond, while Figure 7.16 illustrates the typical pattern you might observe today.

Derelict sites in towns are colonised by plants, and it would be interesting to examine the rate of colonisation of different sites, perhaps relating this to microclimate, size of the site and distance from the nearest source of vegetation. What proportion of plants have seeds dispersed by wind and by birds?

3   A vegetation project with a historical angle is based on the fact that the number of shrub species found in a hedge is a function of the age of the hedge. Observations in many English counties have shown that one new species colonises the hedgerow every one hundred years. Saxon hedges have about ten shrub species, Tudor hedges four and Enclosure Act hedges two. Using this technique you could, for example, date the hedges in an old village and (with the aid of historical records and maps) see how field size and shape have changed over time. Use Figure 7.13 to help to identify species.

### Table 7.3  Field method

1   Choose a section of hedge that appears homogeneous, e.g. does not vary markedly in physiognomy or does not traverse obvious topographical variations.
2   Measure a 30-metre length.
3   Count and record the number and type of tree species.
4   Record the number of different shrub species in the 30-metre length.

### Figure 7.17  Four stages in the evolution of a hydrosere in a freshwater lake

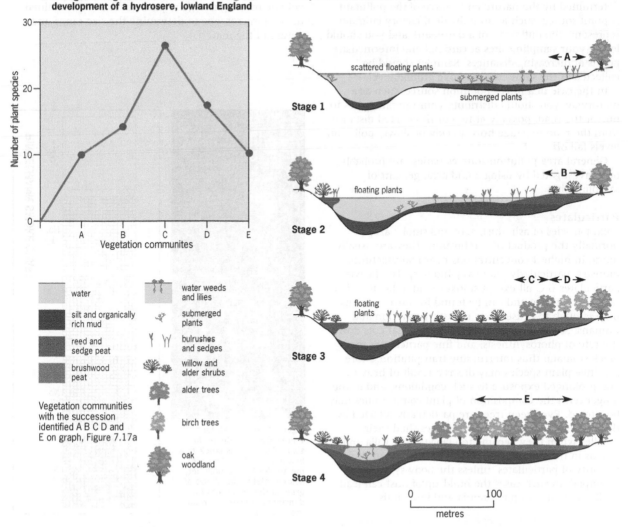

**a  Changes in vegetation communities during the development of a hydrosere, lowland England**

**b  Hydrosere in lowland England**

Vegetation communities with the succession identified A B C D and E on graph, Figure 7.17a

# 7.5 Air pollution

Air pollution has been with us for many centuries – it is not a new phenomenon, and in many respects it is not as serious as it was 50 years ago. London smogs caused by coal burning, for example, are a thing of the past. However, we are now polluting on a far greater scale than before and, while urban air pollution still causes concern (particularly in cities of the developing world), geographers increasingly focus on pollution at the continental scale (acid rain) and the global scale (the greenhouse effect or global warming).

Even though many students are very concerned about the consequences of global warming, this is a wholly impractical scale of study, so the techniques described below relate to local pollution studies that can be tackled by A-level students.

## Sampling

Of greatest importance in all pollution studies is the structure of the sampling frame. This will be determined by the nature and source of the pollutant. A point source, such as an individual factory chimney, represents the bull's-eye of a dartboard, and you should locate your sampling sites at cardinal and intermediate points at increasing distances. Samples should be collected at the same height above ground level.

In the case of a linear pollution source, such as a motorway, you should distribute your sampling sites to mirror the road, possibly at two or three fixed distances from the road to gauge how quickly or slowly pollution levels fall off.

General area pollution sources (cities) are probably best investigated by using a grid arrangement of sampling sites.

## Particulates

Solid particles of ash, dust, soot and smoke are normally the product of combustion. They are usually found in highest concentrations near manufacturing industry, particularly old heavy industry. The heavier particulates (i.e. all except smoke) tend to be deposited near their source and can be found in concentrations high enough to cause harm to both vegetation and humans. A film of dust coating a young leaf cuts down the rate of photosynthesis, and fine particulates may block stomata, thus interrupting transpiration. More sensitive plant species may die as a result of heavy and prolonged exposure to such conditions, and in the longer term the composition of plant communities may be altered. Evergreen trees are particularly affected by this problem, although certain species shed their needles when they become too clogged up. Cilia and mucus in the human respiratory tract filter out the majority of particulates, unless the body systems are swamped. In such cases the build-up of dust can lead to illnesses such as pneumonia and bronchitis.

ADHESIVE STRIPS AND CARDS The simplest and most effective method of measuring the concentration of particulates is to use adhesive strips and cards. These are left in the open, protected from rain, to catch settling dust and smoke. After a given time, compare them under a microscope. Observation should include the range of particle size as well as the number of particles.

A simpler variant is to leave a piece of white card near a partly open window. Place 6–12 wooden blocks on the card and remove one every day. This will give a visual indication of the amount of particulate matter in the air.

FILTER PAPER A filter paper folded into a cone and placed in the mouth of a rain gauge (or a funnel in a milk bottle) collects particulates gathered by rain as it falls through the air. Transfer the filter paper to a numbered plastic bag, dry the paper in an oven and brush the particulates into the pan of a balance to determine the volume by weight. Again, particles may be examined under a microscope to determine the size range and nature of the material.

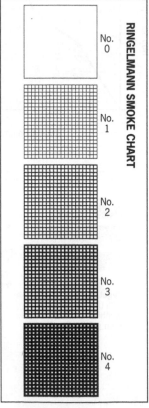

**Figure 7.18 Ringelmann Smoke Chart (reduced in size). The chart is stuck to a board and held in line with the smoke from a polluting source. Match the shade of grey on the chart with the density of smoke and record the number.**

*Smoke*

Smoke from domestic sources was once Britain's most serious air pollutant. However, greater use of gas, oil and electric central heating, along with the Clean Air Acts of 1956 and 1968, have largely eliminated this problem, and changes in industrial production have resulted in fewer smoky chimneys. Smoke emissions can be classified by holding up a Ringelmann Chart and comparing the density of the smoke with the most appropriate of the printed grey scales (Fig. 7.18).

*Lead*

Concentrations of lead in the atmosphere have already begun to decline. This is because permitted levels of lead in petrol are being reduced, and unleaded petrol and diesel-burning cars are becoming more popular. Lead emissions from traffic, which were particularly high near major urban roads, affected the physical and mental development of young children.

Sufficient numbers of cars still use leaded petrol for you to be able to detect atmospheric lead. However, you cannot use the method described below to detect variations in the concentration of lead, simply its presence.

MEASUREMENT OF LEAD IN PLANT TISSUE
Obtain: 50 mg sodium rhodizonate
      deionised water
      1.5 g tartaric acid
      1.9 g sodium bitartrate
      scalpel
      microscope ($\times$ 100–400) and glass slides

*Method*

1 Dissolve 50 mg sodium rhodizonate in 25 ml deionised water.
2 Dissolve 1.5 g tartaric acid and 1.9 g sodium bitartrate in 100 ml deionised water to make a buffer solution.
3 Using the scalpel, obtain very thin slices of plant tissue. Place them in 2 ml sodium rhodizonate solution and leave for 30 minutes. Add a few drops of buffer solution and leave for another 10 minutes.
4 Rinse the plant tissue with deionised water, mount it on a glass slide with the buffer solution and examine under the microscope immediately. Lead is present in the tissue if it turns a bright pink to scarlet colour.

## Gaseous pollutants

Gases are produced by burning and a variety of industrial processes. Measurement of the actual concentration of a gas usually requires expensive apparatus, although you can quite easily detect the presence of a pollutant. Other methods measure the effect of a pollutant on vegetation and work out the concentration of pollutants indirectly.

*Sulphur dioxide*

Sulphur dioxide is one of the most serious air pollutants in the UK. It is emitted principally from coal- and oil-burning power stations, industry and oil refineries, and is a major cause of acid rain. You can measure the concentration of $SO_2$ quite easily by using a Dräger gas monitor or Dräger tubes.

*Carbon monoxide*

Carbon monoxide (CO) is a product of combustion. In towns, therefore, it is produced largely by cars, buses and trucks. Carbon monoxide levels have been rising in many towns because, despite more efficient engines, the sheer number of cars has increased and road congestion has become more severe. We can, to some extent, regard carbon monoxide as a surrogate or proxy indicator for other more harmful forms of vehicular pollution such as lead, nitrogen dioxide, particulates and volatile organic compounds. Care needs to be taken in drawing such comparisons, however: vehicular emissions of lead, for instance, have declined dramatically since the early 1990s and the widespread use of unleaded petrol. Fortunately, carbon monoxide is not toxic although at high concentrations it can reduce the ability of the blood to carry oxygen. The equipment needed to measure carbon monoxide is relatively cheap and quite easy to use.

A CO meter gives a digital reading of carbon monoxide concentrations in parts per million (ppm). The following method is normally used:

1 Hold the meter by the kerbside, 1.5 m high, with the sampling port facing on to the street.
2 Record CO level every 15 seconds over a five-minute period.
3 Calculate the mean of these twenty readings.

The sampling frame within which the CO meter is used needs to be carefully constructed: this will depend upon the aim of your project. It is common to take readings along a transect from city edge to city centre. Alternatively, it can be used to take a cluster of samples within a town and a cluster from outside town. The meter is sensitive and CO concentrations tend to fall sharply away from busy traffic, so it is possible to get interesting variations over short distances – particularly with increasing distance from a busy road.

*Acid rain*

Acid rain is the product of chemical reactions between sulphur dioxide, nitrogen oxides and water vapour. In order to measure the acidity of rain, collect rainwater in a rain gauge (page 84), preferably with a mesh or gauze to keep out leaves and other debris that might react chemically with the water. Test the rainwater for acidity by using narrow-range pH strips (these are more accurate across the pH band of acid rain than strips that measure the full pH range). Alternatively, acid rain test kits are available, or else use a pH meter.

The rain gauge should be rinsed with deionised water between each reading. Rainwater is naturally acidic. This is because carbon dioxide dissolves in rain to produce weak carbonic acid. pH levels in cities tend to be much lower (i.e. the acidity is greater) than in rural areas because of acidic emissions from cars, especially nitrogen oxide and sulphur dioxide.

If you conduct a survey of acid rain, you must take into account the susceptibility of the area to acid rain attack. One of the reasons the Scandinavians complain so strongly about acid rain is because large parts of Norway and Sweden are underlain by acid rock. Also, their abundant coniferous forests add to the acidity of the soil. Consequently, acid rain pushes the pH level of soil and lakes down to dangerously low levels. On the other hand, the calcareous and basic (alkaline) geology of much of southern England neutralises the acidity in the rain so that it does not become a real problem.

In cities, old buildings (even those made of limestone) have been corroded by centuries of acid rain, and the damage is often very obvious. Techniques for investigating the weathering of buildings are described on page 17.

*Ozone*
Ozone has only recently been recognised as a pollutant in the UK. High in the stratosphere, ozone serves a vital function by shielding the Earth from harmful ultraviolet rays. However, at ground level it is harmful to human health and to plants. Ozone is one of a cocktail of pollutants known as photochemical smog. This smog is produced by the chemical reaction of nitrogen oxide (released in car exhausts) with atmospheric oxygen in the presence of bright sunshine. As a result, the effects of ozone tend to be worse in summer than in winter.

Ozone enters leaves through the stomata and attacks the cells, leading to the appearance of brown spots. It also causes the premature ageing and death of leaves (Fig. 7.19).

**Visual assessment of plant damage**
This method is subjective and has lost some of its scientific credibility in recent years. This happened because German forests were supposedly killed by acid rain, but the symptoms allegedly due to acid rain turned out to be nothing more than moisture stress following several years of drought. Visual assessment is a direct measurement of the effect of pollution on vegetation but, as the German example revealed, factors other than air pollution can affect plant health. If you are using this method, you must design your experiment carefully so as to try to isolate the effect of

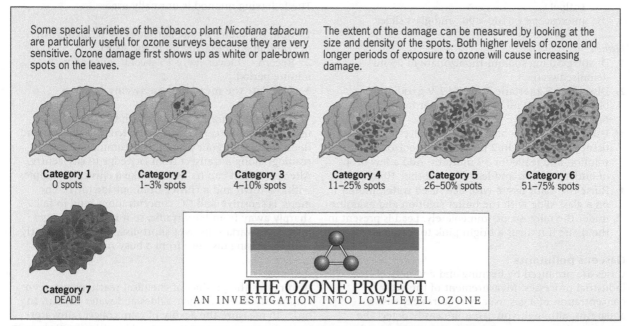

Some special varieties of the tobacco plant *Nicotiana tabacum* are particularly useful for ozone surveys because they are very sensitive. Ozone damage first shows up as white or pale-brown spots on the leaves.

The extent of the damage can be measured by looking at the size and density of the spots. Both higher levels of ozone and longer periods of exposure to ozone will cause increasing damage.

**Category 1** 0 spots

**Category 2** 1–3% spots

**Category 3** 4–10% spots

**Category 4** 11–25% spots

**Category 5** 26–50% spots

**Category 6** 51–75% spots

**Category 7** DEAD!!

**THE OZONE PROJECT**
AN INVESTIGATION INTO LOW-LEVEL OZONE

Figure 7.19 Levels of ozone pollution indicated by categories of damage to the leaves of *Nicotiana tabacum*

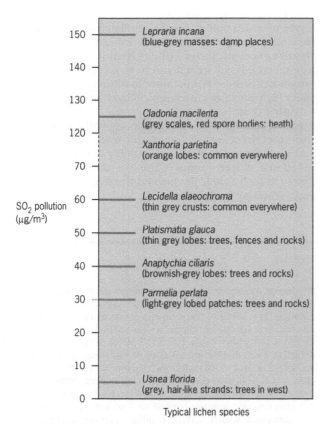

SO₂ pollution (μg/m³)

- 150 — Lepraria incana (blue-grey masses: damp places)
- 140
- 130
- 120 — Cladonia macilenta (grey scales, red spore bodies: heath)
- 70 — Xanthoria parietina (orange lobes: common everywhere)
- 60 — Lecidella elaeochroma (thin grey crusts: common everywhere)
- 50 — Platismatia glauca (thin grey lobes: trees, fences and rocks)
- 40 — Anaptychia ciliaris (brownish-grey lobes: trees and rocks)
- 30 — Parmelia perlata (light-grey lobed patches: trees and rocks)
- 20
- 10
- 0 — Usnea florida (grey, hair-like strands: trees in west)

Typical lichen species

**Figure 7.20  Lichen species as indicators of sulphur dioxide pollution. Lichens are inexact but useful indicators of pollution; they tolerate approximate pollution levels up to those shown in the table.**

pollution. Take into account other possible causes of damage, such as drought, pests, old age and disease. Remember, however, that pollution may have weakened the defences of trees that would not otherwise have been affected by these conditions.

The method is a summer technique and involves observation of the foliage loss of tree crowns. You can use class intervals of: no loss; 5–10 per cent; 11–25 per cent; 26–50 per cent; and more than 51 per cent foliage loss. Also, as some species are more tolerant of polluted environments than others, it is probably better to examine one or two indicator species rather than all the trees present.

### Lichenography

Lichens are very simple plants, but they are highly susceptible to air pollution. This is because lichens have no roots and absorb moisture and solutes through their entire body surface. In polluted environments, concentrations of sulphur dioxide and other pollutants may reach critical levels that slow down growth rates and possibly kill the lichen. However, different lichen species tolerate varying levels of pollution, and we can therefore regard certain species as indicators of the prevailing level of pollution. Figure 7.20 shows tolerance levels of lichen indicator species to sulphur dioxide pollution, and Figure 7.21 will help you with species identification.

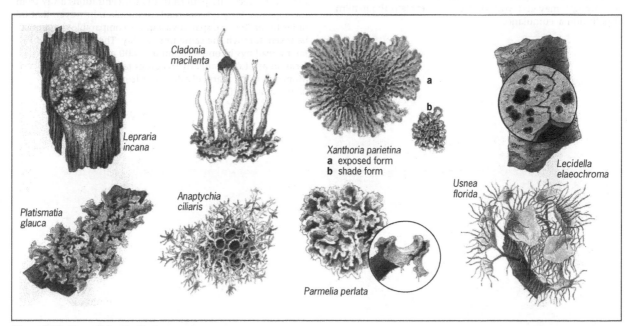

**Figure 7.21  Some lichen indicator species**

### Factors influencing the level of air pollution

A great many geographical variables, apart from the volume of emissions, affect the concentration of pollution. Some of these vary with time, while others (such as topography) remain fixed. It is important to take into account all possible influences on the build-up of pollution.

The most significant influence on the level of air pollution is the state of the atmosphere. Strong winds help to disperse pollution; temperature inversions trap pollution at low levels; precipitation (both rain and snow) cleanses the atmosphere by scavenging – removing solid particles. All of these variables affect the degree to which pollution is accumulated or dispersed. These meteorological variables are influenced by the prevailing synoptic picture (the current state of the atmosphere). It will help if you have access to a barometer (because temperature inversions occur under high-pressure anticyclonic conditions) and to a weather chart.

Season and time of day may affect pollution. An obvious example is rush-hour traffic levels. Less obvious is the variation in susceptibility of plants to pollution at different times of the growing cycle. However, such a topic lies beyond the scope of this book, requiring further research.

## 7.6 Project suggestions: air pollution

Pollution studies must take into account the geography of the area being studied. Do not consider the pollution in isolation, but give thought to conditions that increase or decrease the problem. Also, assess the extent to which ecosystems are able to cope with their pollution loading.

1  Geographical patterns of the concentration and distribution of air pollution make a good study. The aim is to assess the longer-term pattern of the distribution of pollution. Using the bull's-eye sampling frame referred to on page 72, establish a series of monitoring stations around a source of pollution. If you decide to tackle the first of these aims, you will need to record and map pollution data (your choice of techniques will be determined by the type of pollution). Try to obtain climatic data for your nearest meteorological station from the Meteorological Office (page 165). Wind direction and strength may give you some clues to explain the pattern of pollution accumulation.

2  Carbon monoxide makes a good target for studying short-term variations in air quality and variations over short distances. It is better to focus on either one or the other. To investigate temporal variations, set up a sampling frame of locations along busy roads. Use a CO meter to record pollution levels at different times of day to compare morning and evening peak-hour traffic with quieter times. What sort of factors appear to affect CO levels: traffic speed, traffic density? Alternatively, compare pollution levels under different synoptic weather conditions – periods of cyclonic and anticylonic weather. What conditions appear to affect CO levels most – wind speed, temperature, precipitation?

To analyse spatial variations compare CO levels in a town centre, suburban High Street and Green Belt. Or look at variations outside different schools in a town. Alternatively, look at the declining pattern of CO concentrations away from the edge of a busy arterial road. Remember, however, that in order to make these spatial variations comparable they must be taken at much the same time of day. There is no point in taking readings in one location at 09.00 and at another location at 11.00. Similarly, readings taken at 08.00 on a Monday will be markedly different from those taken at 08.00 on a Sunday.

# 7.7 Water pollution

Unlike air pollution, which is predominantly an urban phenomenon, water is badly polluted in the countryside by farm chemicals and in towns by industrial effluent, so it is a problem that can be investigated by rural and urban geographers alike.

Apart from the obvious dangers of working near water, badly polluted rivers are a potentially serious health hazard. You should avoid direct contact with the water as far as possible, particularly if you have a cut. Coliform bacteria from sewage and Weil's disease (spread by rats) are real dangers. During hot weather, blooms of blue-green algae may appear as a scum on the water surface. This can cause severe irritation of the skin – it must be avoided. Wash your hands thoroughly before eating.

There are so many forms of freshwater pollution that you can use several of your senses to detect them. For example, the colour of the water, the presence of foam, oil and dead fish are tell-tale signs, and a badly polluted watercourse will smell. Raw sewage is still poured into some rivers in the UK. Rivers that have no dissolved oxygen are anaerobic and often give off methane or vile-smelling gases such as hydrogen sulphide and ammonia.

Just because a river bears none of these outward signs of pollution does not necessarily mean that it is clean, and a little scientific detective work may be needed to establish its precise level of contamination. The techniques described here deal with organic pollution only: inorganic pollutants such as heavy metals require apparatus that is not normally available in schools and colleges.

## Dissolved oxygen

Oxygen dissolved in water is required by most forms of aquatic (water) life. A healthy river is a well-aerated river, one in which the biological demands for oxygen are smaller than the supply. Oxygen is produced by photosynthesising river plants and is absorbed from the atmosphere, particularly by fast-flowing and turbulent streams.

Micro-organisms such as bacteria and blue-green algae multiply rapidly under favourable conditions. They have such a great appetite for dissolved oxygen that, in badly contaminated rivers, dissolved oxygen falls to dangerously low levels. Low oxygen levels threaten the life of normal aerobic (oxygen-demanding) forms of aquatic life. This most commonly occurs downstream of sites where untreated sewage, farmyard manure, slurry and organic industrial waste are poured into rivers.

### Eutrophication

Enrichment of water by nutrients is a process known as eutrophication. It is one of the most serious of water pollution problems in the UK today. It is caused by the discharge of raw sewage (with its associated oxygen-demanding bacteria), but is particularly bad in areas of intensive agriculture when chemical fertilisers (notably nitrates) are washed off the land or are leached through the soil into ditches, streams and ponds. Here the fertilisers stimulate the growth of algae and other freshwater organisms that remove oxygen from the water. Aquatic plants may choke the river and slow the rate of flow to such an extent that oxygenation from the atmosphere is reduced.

### Measurement of dissolved oxygen

This is best achieved with an electronic meter. The exact method will depend upon the type of meter you have. In all cases the instrument has a probe that can either be inserted directly into the water or into a water sample taken from the river. In both cases you must take care to avoid creating air bubbles, which would distort the reading.

As oxygen becomes less soluble at higher temperatures, it is important to calibrate your reading against oxygen-saturated water at the same temperature. To do this, take a half-full beaker of water and shake it for a minute, then set the meter to read 100 per cent in this sample.

### Biological oxygen demand

Biological oxygen demand (BOD) is a useful indication of bacteriological levels in streams and, therefore, pollution.

1  Collect 500 ml of water in two sealable bottles; determine the dissolved oxygen content of both.
2  Seal both bottles and wrap one in aluminium foil to prevent sunlight from reaching it.
3  Leave both for 2–3 hours.
4  Measure the dissolved oxygen content of both bottles. The difference represents the BOD.

In the bottle exposed to light, photosynthesis as well as respiration will have taken place, whereas in the darkened bottle, respiration only will have occurred.

## Nitrates

As with many other pollutants, a certain amount of nitrate is essential for plant growth. Problems occur when concentrations of nitrates rise above a certain level. EU regulations specify a maximum concentration of 50 mg/l in drinking water, although this level has been challenged as being entirely arbitrary: scientists say there is no proven harm to health if concentrations exceed this limit. Nevertheless, the water authorities are working hard to reduce nitrate levels in rivers.

Nitrates can be measured in the soil, in plants and in water. The source of the nitrate in soil is largely from chemical nitrate fertilisers, and nitrates in water have been leached from soil or come from sewage.

Nitrates can be detected in several ways, although the use of nitrate test strips is the cheapest and simplest method and can be done in the field. Dip the strip into water and compare the resultant colour with a chart to obtain the nitrate concentration in milligrams per litre. Nitrates can also be measured with electronic probes. Water samples can be taken back to the laboratory for analysis, but analysis must take place within 1–2 hours.

## Calefaction

Calefaction is heat pollution. Although warmth alone may not be as much of a threat as certain chemicals, it is still worth investigating, particularly near the most significant source of heat, namely power stations. At low river discharge, temperature increases of up to 8 °C have been recorded downstream of power stations. Oxygen becomes less soluble in water at higher temperatures, and the temperature rise could reduce oxygen to dangerously low levels in an already polluted river. Higher temperatures can prevent the spawning of certain types of fish and slow down the development of the eggs. Water temperature may vary during the day, so you should take temperature readings at similar times each day.

## Turbidity

Turbidity describes the opaqueness of water, which is a natural phenomenon (because of suspended sediment) but can also be a measure of pollution. You can measure it using a turbidity meter (a photoelectric cell set 1 cm from an electric light source, both mounted on a probe immersed in the water) or more simply by using a Secchi disc.

The Secchi disc is lowered into the water until it cannot be seen, and then raised until it has come back into view. You record the depth at which it disappears from view and compare this at different sites. Factors other than turbidity can influence readings, in particular the amount of sunlight, shade from overhanging trees, time of day and so on.

## pH

The acidity of water can be measured by using narrow-range test strips or a pH probe. Like rainwater, river water is often naturally slightly acidic (pH below 7.0), particularly in areas of acidic geology – parts of Devon and Cornwall, large parts of Wales, the Lake District and most of Scotland – so a pH below 7.0 does not necessarily indicate pollution. Much also depends on the time of year: a melting snow pack isolates and concentrates acid, and when it finally melts a burst of low pH meltwater is flushed into upland streams. This can be immensely damaging, particularly if the snow

has been around for a long time accumulating acid precipitation.

Needless to say, abrupt downstream decreases in pH or low pH levels downstream of industrial outflows or cities have different explanations and should be investigated further.

## Hydrogen sulphide

Apart from the unpleasant (rotten eggs) smell of this gas, it can be detected chemically.

1   Collect 500 ml river water in a 1 litre bottle.
2   Add a few drops of any acid and place a strip of lead acetate paper in the neck of the bottle.
3   The paper turns black if hydrogen sulphide is present. This indicates the presence of anaerobic bacteria. At very low or zero oxygen levels these bacteria replace oxygen-breathing bacteria. Such rivers are very badly polluted and may be almost dead.

## Ammonia

Another gas produced by anaerobic bacteria is ammonia. You can detect it by placing a drop of river water on to turmeric paper, which turns from yellow to brown if ammonia is present. Ammonia is not only the product of the decomposition of organic wastes; it is also an important constituent of farmyard manure and untreated sewage. pH values of river water will rise above 7.0 if ammonia (an alkali) is present in quantity.

## Indicator species

The concept of indicator species (plants or animals) in air pollution studies was mentioned on page 75. The same principle can be applied to freshwater studies.

Use a net to collect the fauna. A seaside net will do, although scientific nets with one flat side for sweeping the stream bed can be bought. You should spend 5–10 minutes collecting at each site to ensure that you have sampled each of the different habitats in a stream (under boulders, in the mud, under the banks, among weed). In a gravelly channel you will need to kick-sample: hold the net at arm's length downstream while you disturb the bed with your feet for 30 seconds. The current will carry the fauna into the net.

Tip the animals into a white-bottomed dish and try to identify and count the species present. Figure 7.22 illustrates and describes some of the more important freshwater invertebrates. The population profile gives you an idea of water quality.

Fish species also have different tolerance levels of pollution. Salmon are well known for avoiding all but the cleanest of rivers, which explains the importance of salmon returning to the Thames. It would be impractical for you to survey fish types yourself, but you can interview anglers on the bank and ask them

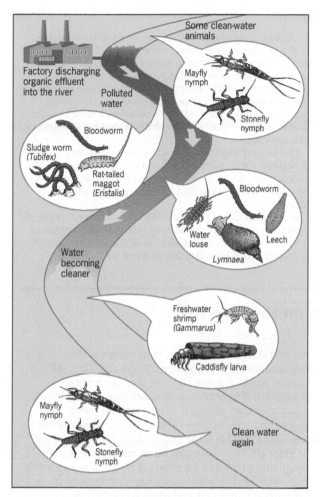

**Figure 7.22 Freshwater fauna as indicators of river pollution. This figure can be used in the identification of stream invertebrates.**

what they have caught over the previous year. Table 7.4 illustrates some results from a survey carried out by Claudia Harrison.

### Factors influencing water-pollution levels

As with air pollution, factors other than the quantity of pollutant determine the recorded pollution levels of rivers. In particular, the ability of the stream to clean itself needs to be investigated. From the earlier section it should be obvious that dissolved oxygen levels are critical: two similar-sized streams given the same volume of untreated sewage will show up very differently if one stream aerates itself with faster flow and turbulence.

Variables such as discharge and velocity (see Section 4.3) must be recorded. Width-to-depth ratios may be important too. Seasonal changes will be considerable: the number of animals will change between summer and winter; discharge will fluctuate; river temperature will affect dissolved oxygen levels; dead leaves in the river will reduce oxygen levels; recent rain and flooding will affect suspended sediment concentrations and turbidity. All these influences must be monitored and taken into account when you are analysing patterns of pollution.

Similarly, you must take care with the construction of the sampling frame. The source of pollution may be a single point (a sewage outfall) or it may be linear (nitrates seeping from many hectares of farmland). Most pollution studies examine the impact of a particular source, and you should use a 'before and after' (or rather 'above and below') type of sampling pattern.

One difficulty for environmentalists when they try to collect evidence against polluters is the episodic nature of effluent release. Many polluters store pollution during the day and release it at night when it cannot be seen. Unfortunately, it is not possible to monitor the river continuously, but by sampling at frequent intervals it should be possible to detect most pollution.

**Table 7.4 Results of an angler survey using fish as indicators of river pollution (*Source:* Harrison, 1985)**

**Key**

H   = Fish preferring a high oxygen content
HM = Fish preferring a high–medium oxygen content
M   = Fish preferring a medium oxygen content
ML = Fish preferring a medium–low oxygen content

| Site | Species | | | | | | | | |
|---|---|---|---|---|---|---|---|---|---|
| | Salmon H | Chub HM | Bleak HM | Dace HM | Roach ML | Perch M | Bream ML | Gudgeon HM | Pike ML |
| Oxford | | ✓ | ✓ | ✓ | ✓ | ✓ | | ✓ | ✓ |
| Caversham | | | ✓ | | ✓ | ✓ | ✓ | ✓ | ✓ |
| Henley | | | | | | ✓ | | | ✓ |
| Maidenhead | | | | | ✓ | | ✓ | | ✓ |
| Windsor | ✓ | | ✓ | | ✓ | ✓ | | | ✓ |
| Staines | | | | | ✓ | ✓ | ✓ | | ✓ |

## 7.8 Project suggestions: water pollution

**1** Suggestions for a cost–benefit analysis approach to air pollution were given on page 76, and you can adapt and use these ideas for water pollution. You would first need to assess the full extent of the pollution (the removal of which represents the 'benefit', to be offset against the 'cost'). Carry out a careful survey of aquatic wildlife above and below the source of pollution. Conduct a questionnaire of attitudes among any nearby residents: How do they perceive the environmental quality of the two areas? Write to the manager or owners of the factory concerned; ask them what the cost would be of installing machinery to reduce the pollution.

What would be the impact on the local economy if the factory closed? How many people would lose their jobs? What would be the wider economic and social costs? Does the benefit justify the cost of stopping the pollution? Which other groups of people would agree with your view or oppose it?

**2** Examine the impact of a factory or an industrial estate, or a large pig or cattle farm, on the quality of river water. To do this you will first need to sample water quality upstream of the pollution source. Measure water quality as near as you can to the source(s) of pollution and then at regular intervals downstream. Depending on the severity of the pollution and whether any more pollution is added, water quality may eventually return to the same level that you first recorded.

One of your objectives should be to see how far downstream the effect of the pollution can be detected. You will probably find that one measure of pollution (perhaps BOD) recovers quicker than another (perhaps aquatic fauna).

Some measures of pollution will vary quite a lot, perhaps even from day to day. Measure pollution levels at different discharge levels. Increasing river flow, for example, would be expected to have a diluting effect on the pollution.

## 7.9 Ecology and pollution data sources on the internet

### Ecology
Natural England, or for Wales the Countryside Council for Wales. Through these sites you can order maps of Sites of Special Scientific Interest or download digital maps for use in a Geographical Information System. Through the home page of the Wildlife Trusts you can find links to many sites of interest.

### Pollution
An excellent place to begin your search is the Department for Environment, Food and Rural Affairs (Defra) website. From this site you can access data records for different air pollutants including sulphur dioxide, ozone, nitrogen oxides, carbon monoxide and particulates at over 60 locations around the UK. Data are available in hourly or daily form, tabulated or graphed. There are also interactive databases on air quality providing an amazing range of information. The www.environment-agency.gov.uk website gives information about air pollution, coastal erosion, flooding and water pollution.

# $8$ Local climate

At the local scale, individual land uses may affect atmospheric conditions to such an extent that local climate is modified. Such modifications can be quite noticeable and are usually confined to the spatial limits of the land use. These climatic modifications are known as microclimates, and their study lies well within the scope of an A-level project.

The basis of a microclimate project is the observation and recording of weather. You must measure some or all of the elements of meteorology. As the following section reveals, there is often more than one way of measuring a single variable. The exact method you use will reflect your scientific objective as well as the tools available.

## 8.1 Radiation and sunshine

Solar radiation (insolation) lies towards the short-wave end of the electromagnetic spectrum, much of it in the visible band. It is usually an important influence on temperature. Unless you have access to a radiation recorder (which can record incident (incoming) as well as reflected (outgoing) radiation), the most scientific way of measuring radiation is with a camera light meter. Set the aperture at $f5.6$, or thereabouts, and record the indicated shutter speed; a reading of 1/60 indicates that there is roughly half the amount of light as a reading of 1/125.

You must be very careful in your method, as the camera will record reflected light as well as incident light. Pointing the camera at a dark surface and then at a light-coloured surface will give vastly different readings, even if the amount of incoming radiation

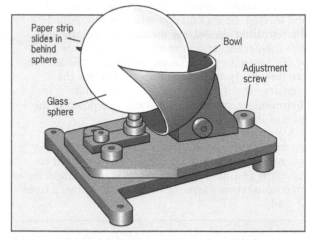

**Figure 8.1  Campbell–Stokes sunshine recorder**

remains exactly the same. Point the camera at a background of standard colour and density.

### Sunshine hours
A Campbell–Stokes sunshine recorder such as that illustrated in Figure 8.1 acts like a giant magnifying glass. It concentrates the sun's rays so that they scorch a line on a strip of paper. As the sun moves across the sky, the scorch line grows. If the sun is blotted out by cloud, the scorching stops. Because the paper strip is scaled in hours, you can calculate the number of sunshine hours in a day, and this, in turn, will help explain variations in temperature. The Meteorological Office also uses Kipp amd Zonen sunshine duration sensors which are electronic.

## 8.2 Temperature

When meteorologists talk about temperature, they are referring to shade temperatures, and unless they specify otherwise they are talking about temperatures measured 1.25 m above the ground. Digital thermometers linked to a data logger will allow you to record continuously over a long period.

### Thermometers
A widely used thermometer is Six's max–min thermometer shown in Figure 8.2. This reads the current temperature and records the maximum and minimum temperatures reached since it was last reset. This is important if you cannot monitor the site constantly. It is particularly useful for recording overnight lows.

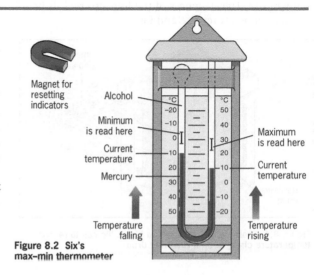

**Figure 8.2  Six's max–min thermometer**

The instrument consists of a U-tube with an alcohol reservoir at one end and a column of mercury occupying the U. As the temperature rises, the alcohol expands and shunts the mercury, which pushes a small iron marker up the max column. As the temperature falls, the alcohol contracts, drawing the mercury back and leaving the max marker where it was. Pulled by the retreating alcohol, the mercury pushes the min iron marker higher up the min column, where it will be left when the temperature rises again. The alcohol can flow freely around the min marker, but the mercury cannot. To reset the Six's max–min thermometer, stroke the metal markers down to the level of the mercury with a magnet.

A *grass minimum thermometer* is used to record overnight lows (Fig. 8.3). It is placed on the ground, where temperatures are usually lower. It is used for recording ground frosts and for measuring temperature inversions when a layer of warm air sits above a layer of cold.

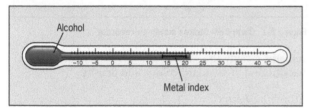

**Figure 8.3 Grass minimum thermometer**

## Thermograph

The thermograph is an instrument that constantly monitors temperature (Fig. 8.4). It operates by using the principle of the bimetallic strip. This strip is made of two metals with different coefficients of expansion fused together, which bends when heated. The strip is attached to a pen that moves up and down in response to temperature changes, plotting a trace on to a chart fixed to a revolving drum. The trace records changes in temperature. The steepness of the line also reveals how quickly temperatures rise and fall.

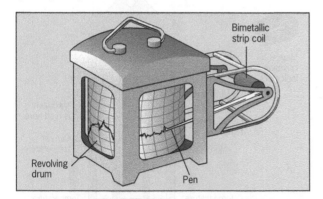

**Figure 8.4 Thermograph that uses a bimetallic coil to record temperature changes on a revolving drum**

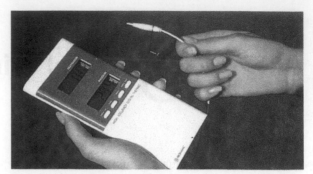

**Figure 8.5 Electronic digital thermometer**

## Electronic thermometers

Cheap electronic thermometers are now readily available in schools (Fig. 8.5). They are usually max–min thermometers and have advantages and drawbacks over conventional alcohol and mercury instruments. The biggest supposed advantage is their precision: they show the temperature in tenths of degrees, which is marvellous in many microclimate studies where it may not be possible to measure small variations in temperature with conventional thermometers. However, the manufacturers' claims can be misleading: they will quote the accuracy as ± 0.2 °C, but in practice two identical instruments lying side by side may show a difference of as much as 1 °C.

A greater drawback is the fact that electronic thermometers do not work when they get wet. Although the instruments normally recover once they are dry, data may be lost. Ideally, they have to be mounted in waterproof housings. Polythene bags will not serve this purpose, as they create a mini greenhouse effect. Some electronic thermometers have a probe attached: this is waterproof and, provided the electronic box is shielded from sunlight, the instrument should work.

## Calibration

If you are using more than one thermometer, then they must all be calibrated before you take any readings. This is because even identical thermometers may give slightly different readings, which in microclimate studies may confuse your findings. Number all thermometers first. Then place them either in cold water (do not immerse electronic thermometers) or in a fridge. Read and record all temperatures. Then immerse the thermometers in warm water and record their readings. Any differences should be consistent over a range of temperatures, and you can adjust the readings to a norm.

## Stevenson screen

If readings are to be taken over a long period, it is worth buying or making a Stevenson screen. This is a louvred box made of white wooden slats and mounted on legs. In it, at a height of 1.25 m, are housed thermometers and other instruments. Its chief function is to provide a shaded and ventilated environment and uniform conditions in order to compare sites. Even if you do not use a Stevenson screen, place thermometers at a standard height in shaded areas.

# 8.3 Humidity

Humidity is the amount of water vapour in the air. This can be expressed in several different ways, but one of the most common is relative humidity. This describes, as a percentage, the amount of moisture in the air in relation to the maximum amount the air could hold at that temperature.

## Wet and dry bulb hygrometer

Two identical thermometers lie side by side. The mercury bulb of one, the wet bulb, has a muslin wick over it that is connected to a small water supply. The other thermometer is the dry bulb (Fig. 8.6). Water evaporates from the wet bulb, and the loss of energy involved causes the wet bulb to read a lower temperature. Record the temperatures of both bulbs, and refer to a conversion chart to determine relative humidity as a percentage (Table 8.1). In dry air, evaporation from the wet bulb is rapid, and so the temperature difference will be greater. In saturated air no evaporation occurs, so the dry and wet bulbs record the same temperature.

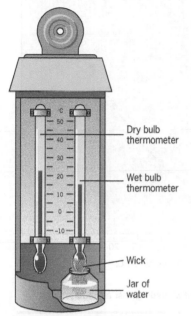

**Figure 8.6 Wet and dry bulb hygrometer**

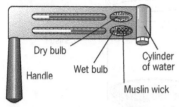

**Figure 8.7 Whirling psychrometer**

## Whirling psychrometer

This operates by using exactly the same principle as wet and dry bulbs, but is a more mobile instrument. A wet and dry bulb lie side by side in an instrument that looks like a football rattle (Fig. 8.7). Whirl it at shoulder level for about a minute, stop and read the wet bulb. Continue whirling until the wet-bulb temperature has stopped falling. Then read off the humidity by using a conversion chart.

**Table 8.1  Tables used for converting wet and dry bulb readings into relative humidity (%)**

| Dry Bulb Temp °C | Depression of the wet bulb (°C) | | | | | | | | | | | | | | | | |
|---|---|---|---|---|---|---|---|---|---|---|---|---|---|---|---|---|---|
| | 0.5 | 1 | 1.5 | 2 | 2.5 | 3 | 3.5 | 4 | 4.5 | 5 | 5.5 | 6 | 6.5 | 7 | 7.5 | 8 | 8.5 |
| 30 | 96 | 93 | 89 | 86 | 83 | 79 | 76 | 73 | 70 | 67 | 64 | 61 | 58 | 55 | 52 | 50 | 47 |
| 29 | 96 | 93 | 89 | 86 | 82 | 79 | 76 | 72 | 69 | 66 | 63 | 60 | 57 | 54 | 52 | 49 | 46 |
| 28 | 96 | 93 | 89 | 85 | 82 | 79 | 75 | 72 | 69 | 65 | 62 | 59 | 56 | 53 | 51 | 48 | 45 |
| 27 | 96 | 92 | 89 | 85 | 82 | 78 | 75 | 71 | 68 | 65 | 62 | 59 | 55 | 52 | 50 | 47 | 44 |
| 26 | 96 | 92 | 88 | 85 | 81 | 78 | 74 | 71 | 67 | 64 | 61 | 58 | 55 | 51 | 49 | 46 | 43 |
| 25 | 96 | 92 | 88 | 84 | 81 | 77 | 74 | 70 | 67 | 63 | 60 | 57 | 54 | 50 | 47 | 44 | 41 |
| 24 | 96 | 92 | 88 | 84 | 80 | 77 | 73 | 69 | 66 | 62 | 59 | 56 | 52 | 49 | 46 | 43 | 40 |
| 23 | 96 | 92 | 88 | 84 | 80 | 76 | 72 | 69 | 65 | 62 | 58 | 55 | 51 | 48 | 45 | 42 | 39 |
| 22 | 96 | 92 | 87 | 83 | 79 | 76 | 72 | 68 | 64 | 61 | 57 | 54 | 50 | 47 | 44 | 40 | 37 |
| 21 | 96 | 91 | 87 | 83 | 79 | 75 | 71 | 67 | 63 | 60 | 56 | 52 | 49 | 46 | 42 | 39 | 35 |
| 20 | 96 | 91 | 87 | 83 | 78 | 74 | 70 | 66 | 62 | 59 | 55 | 51 | 48 | 44 | 41 | 37 | 34 |
| 19 | 95 | 91 | 86 | 82 | 78 | 74 | 70 | 65 | 61 | 58 | 54 | 50 | 46 | 43 | 39 | 35 | 32 |
| 18 | 95 | 91 | 86 | 82 | 77 | 73 | 69 | 65 | 60 | 56 | 52 | 49 | 45 | 41 | 37 | 34 | 30 |
| 17 | 95 | 90 | 86 | 81 | 77 | 72 | 68 | 64 | 59 | 55 | 51 | 47 | 43 | 39 | 35 | 32 | 28 |
| 16 | 95 | 90 | 85 | 81 | 76 | 71 | 67 | 62 | 58 | 54 | 50 | 46 | 41 | 37 | 34 | 30 | 26 |
| 15 | 95 | 90 | 85 | 80 | 75 | 71 | 66 | 61 | 57 | 52 | 48 | 44 | 40 | 36 | 31 | 27 | 24 |
| 14 | 95 | 90 | 84 | 79 | 74 | 70 | 65 | 60 | 56 | 51 | 47 | 42 | 38 | 33 | 29 | 25 | 21 |
| 13 | 95 | 89 | 84 | 79 | 74 | 69 | 64 | 59 | 54 | 49 | 45 | 40 | 36 | 31 | 27 | 23 | 18 |
| 12 | 94 | 89 | 83 | 78 | 73 | 68 | 63 | 57 | 53 | 48 | 43 | 38 | 34 | 29 | 24 | 20 | 16 |
| 11 | 94 | 88 | 83 | 77 | 72 | 66 | 61 | 56 | 51 | 46 | 41 | 36 | 31 | 26 | 22 | 17 | 13 |
| 10 | 94 | 88 | 82 | 77 | 71 | 65 | 60 | 54 | 49 | 44 | 39 | 34 | 29 | 24 | 19 | 14 | 09 |
| 9 | 94 | 88 | 82 | 76 | 70 | 64 | 58 | 53 | 47 | 42 | 36 | 31 | 26 | 21 | 16 | 11 | 06 |
| 8 | 94 | 87 | 81 | 75 | 69 | 63 | 57 | 51 | 45 | 40 | 34 | 29 | 23 | 18 | 12 | 07 | 02 |
| 7 | 93 | 87 | 80 | 74 | 67 | 61 | 55 | 49 | 43 | 37 | 31 | 26 | 20 | 14 | 08 | 02 | 00 |
| 6 | 93 | 86 | 79 | 73 | 66 | 60 | 53 | 47 | 41 | 35 | 29 | 23 | 17 | 11 | 05 | 00 | 00 |
| 5 | 93 | 86 | 79 | 72 | 65 | 58 | 51 | 45 | 38 | 32 | 26 | 20 | 14 | 08 | 02 | 00 | 00 |
| 4 | 92 | 85 | 78 | 70 | 63 | 56 | 49 | 42 | 35 | 28 | 21 | 14 | 07 | 00 | 00 | 00 | 00 |
| 3 | 92 | 84 | 77 | 69 | 62 | 54 | 47 | 40 | 32 | 25 | 17 | 10 | 03 | 00 | 00 | 00 | 00 |
| 2 | 92 | 84 | 76 | 68 | 60 | 52 | 44 | 36 | 28 | 20 | 12 | 04 | 00 | 00 | 00 | 00 | 00 |
| 1 | 91 | 83 | 75 | 67 | 58 | 50 | 41 | 32 | 24 | 15 | 07 | 00 | 00 | 00 | 00 | 00 | 00 |

To calculate relative humidity, find the dry bulb temperature down the left-hand column; the difference between the wet and dry bulb readings is the depression of the wet bulb. Where the dry bulb temperature crosses the depression of the wet bulb, humidity is read as a percentage. For example, dry bulb = 24 °C, wet bulb = 20 °C, depression of the wet bulb = 4 °C, RH = 69%.

Relative humidity is mainly controlled by temperature. As air temperature rises, so does its capacity for holding water vapour (with constant absolute humidity there is a close inverse relationship between relative humidity and air temperature). It is therefore possible to record different relative humidities even though there has been no change in the amount of water vapour in the air, only a change of temperature.

Many meteorologists therefore prefer to measure absolute humidity (grams of water vapour per m³) rather than relative humidity, which often just rises and falls with temperature.

To do this, record relative humidity, then refer to Table 8.2 or Figure 8.8 which show absolute humidity – in other words, the maximum amount of water vapour the air can hold at different temperatures. For example, at 20 °C, saturation point is 17.3 g/m³. If you record 70 per cent relative humidity at a dry-bulb temperature of 20 °C, then absolute humidity is 17.3 × 70/100 = 12.11 g/m³.

| Table 8.2 The relationship between temperature and absolute humidity (*Source*: Gates, 1972) | Temperature °C | Absolute humidity moisture content gm³ |
|---|---|---|
| | –15 | 1.6 |
| | –10 | 2.3 |
| | –5 | 3.4 |
| | 0 | 4.8 |
| | 5 | 6.8 |
| | 10 | 9.4 |
| | 15 | 12.8 |
| | 20 | 17.3 |
| | 25 | 22.9 |
| | 30 | 30.3 |
| | 35 | 39.6 |
| | 40 | 50.6 |

*Hygrograph*

Continuous recording of humidity is possible with a hygrograph. This is an instrument that uses the fact that horse hairs expand when wet and contract when dry. Such is the sensitivity of horse hairs to changes in moisture that a pen connected to them through a system of levers can plot the changes in atmospheric humidity on to a chart on a revolving drum. The units are usually arbitrary.

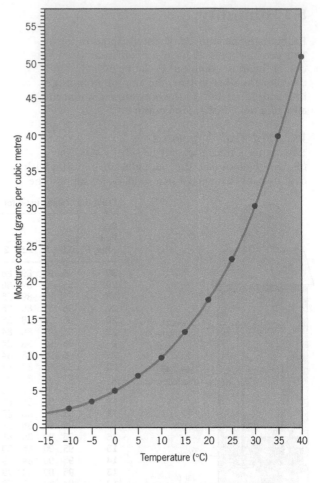

**Figure 8.8  Moisture content plotted against air temperature**

## 8.4 Rain

Rainfall is measured in a rain gauge (Fig. 8.9). This can be home-made (a funnel and milk bottle) or can be purchased cheaply. If you are using home-made gauges, make sure they all have the same diameter aperture and are recorded in the same measuring cylinder. Differences in the diameter of the receiving funnel and measuring cylinder will obviously produce inaccurate results.

Site your rain gauges carefully. You should not place them too near a building or vegetation (unless your aim is to record interception by trees) and you should not leave them in positions that are too exposed. Rain gauges that stick up above the ground tend to underestimate the amount of rain, as air currents and eddies created by the gauge cause rain to be carried past. For this reason, water authorities dig their gauges into the ground, so that the aperture is flush with ground level. You need not do this, provided each gauge is set in a similar position.

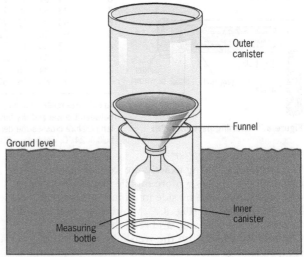

**Figure 8.9  Non-recording rain gauge**

Recording rain gauges can be purchased, and these use a mechanism, such as a float or a tipping bucket, to monitor rainfall. The advantage of float-type gauges is that they can record the number of individual rain events, rain intensity and duration as well as total amount.

Snow can be measured, but not in a rain gauge (which the snow will, of course, block). Measure the depth of snow with a ruler and melt samples of known volume to find the water equivalent.

## 8.5 Atmospheric pressure

Air pressure is measured with a barometer. Aneroid barometers work on the principle of a partially evacuated collapsible cylinder that rises and falls in response to changing pressure. This movement is magnified by a system of levers, and either drives a needle that points to the appropriate part of a dial (Fig. 8.10) or, in a barograph, is attached to a pen that records a trace on a chart on a revolving drum (Fig. 8.11).

A mercury barometer relies on air pressure to support a level of mercury in an evacuated glass tube: rising air pressure pushes the mercury higher up the tube, and this is shown in units of pressure (millibars).

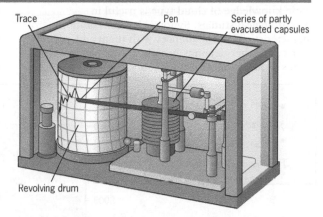

Figure 8.11 A barograph, which records an atmospheric pressure trace on a revolving chart

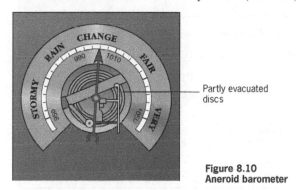

**Figure 8.10
Aneroid barometer**

Pressure is important, because it helps us to understand changes in the synoptic weather situation. Falling pressure warns us that a depression is coming (with its associated fronts, wind and rain), while a rising glass tells us that we shall soon experience an anticyclone or ridge of high pressure. Pressure alone, however, is not an infallible guide to meteorological change and should be used together with weather charts (see page 86).

## 8.6 Wind

Air rushes from place to place as a result of differences in atmospheric pressure. Both wind speed and direction are important for meteorological studies, particularly microclimate. Wind strength may be such that heat islands (local positive-temperature anomalies) are reduced or disappear altogether, while wind from one quarter will bring temperature and humidity characteristics very different from those you experience when another air mass affects you.

### Anemometer
Wind is normally measured with a revolving cup anemometer (Fig. 8.12). The faster the wind, the more quickly the anemometer revolves. Speeds are measured either directly on a scale (in metres per second, knots, miles per hour, kilometers per hour or the Beaufort scale), or a counter shows the number of revolutions, which you can then convert to speed by using a table.

Figure 8.12 A revolving cup anemometer in use

Wind speed is the most variable of all weather phenomena and can change from minute to minute. Observers often make many recordings over a period of several minutes in order to obtain representative readings. Some hand-held anemometers integrate wind speed over a 30-second period, whereas others record maximum gust speed. Both measurements are valid, but be consistent.

## 8.7 Cloud

Some knowledge of cloud type is useful in meteorological studies. Synoptic features bring characteristic cloud sequences with them and, of course, weather depends heavily on the pattern of clouds above. A meteorologist should be able to identify the major cloud types (Fig. 8.13), and he or she should also be able to estimate the amount of cloud cover. By convention, cloud cover is expressed as the number of eighths (or oktas) of sky covered by cloud.

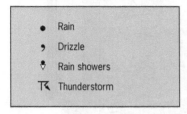

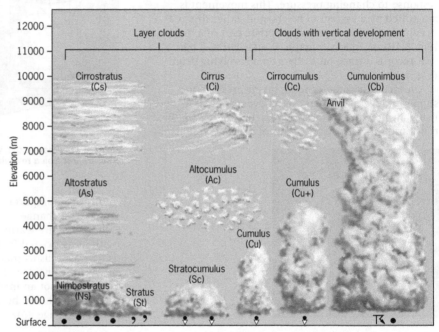

**Figure 8.13 Guide to the identification of major cloud types**

## 8.8 Meteorological observations

The frequency of weather readings will depend upon the scientific purpose of the project. Most micrometeorological studies are carried out over 5–7 days (not necessarily continuously, and preferably under different synoptic conditions) and observations are made every 4 hours during the day, with readings taken as early in the morning and as late at night as possible. When you take early-morning readings, you should read overnight minimum temperatures and reset your thermometers.

The distribution of stations is important. One of the better tried and tested project ideas is to examine the microclimate of a wood or built-up area. To do this you need a small cluster of stations (3–5) in the woodland (or town) and slightly fewer in a nearby open area to act as a control. The temperatures you record are samples, and more samples produce more reliable results. This arrangement also enables you to do tests of significance. Generally, your instruments will not be precise enough to allow you to examine climatic gradients along a transect.

### Synoptic charts

Synoptic charts show the current (or forecast) state of the atmosphere at a given moment. The pattern of isobars illustrates the main synoptic features (highs, lows, ridges and troughs) and symbols show the weather experienced at different sites. Although you will not do a project on anything other than a very local scale, it is important to explain the local conditions you experience in terms of the broader synoptic picture. For example, local sea breezes will be felt only under generally calm (anticyclonic) conditions and will be replaced by regional winds when cyclonic conditions prevail.

Synoptic charts with isobars and fronts are printed in most newspapers, and you should cut out and keep these for the duration of your project. Alternatively, for the most up-to-date synoptic picture log on to the Meteorological Office Website. If your college or school has a weather satellite receiver, you can tie in local conditions and the synoptic chart with what you see on the screen.

### Remote sensing satellites

There are two main types of satellites. *Communications satellites* are used to bounce television, radio and telephone signals from one part of the globe to another. The receiving dishes you see on houses gather pictures from communications satellites.

*Remote sensing satellites* take pictures of the earth and transmit them. There are two main types of remote sensing satellites: *weather satellites* which record cloud patterns and help with weather forecasting, and satellites recording *land use* such as Landsat and the French SPOT satellite.

There are two main types of weather satellite: *polar orbiting* and *geostationary*. Some NOAA (National Ocean and Atmospheric Administration) satellites are polar orbiting, which means that they orbit from pole to pole with the Earth spinning beneath them. Because the polar-orbiting satellites are at a height of only 540 miles, they are able to transmit detailed images. Images are available on the Meteorological Office website.

The second form of weather satellite, such as Meteosat, is geostationary. It is in a fixed position 22 240 miles above the Earth's surface and spins with the Earth. It transmits a picture of a different part of the Earth every four minutes.

Remote sensing satellites are able to receive and transmit two types of image, visible and infra-red. The *visible image* is a measure of the amount of light reflected back by clouds and the Earth's surface, while *infra-red images* indicate temperature: the colder the surface, the whiter it will appear.

### Climatic data

Some projects require the use of longer-term climatic data. Investigations of longshore drift, for example, will probably need to look at the longer-term pattern of wind direction and strength. In this case you will need to use secondary data compiled by the Meteorological Office, which collects and processes data from nearly 300 monitoring stations around the UK and from 5000 rain gauges. It should therefore be possible to obtain data from a site not too far from you.

## 8.9 Project suggestions

Just about all land uses create their own microclimate, but the land use has to be distinctive enough and large enough to create a measurable impact. Towns, woods and lakes are often studied. The effect of topography, the influence of a hillside or valley, is less straightforward, but workable.

As a general starting point, a microclimatic project could aim to examine the intensity of the microclimate under different synoptic conditions. This means either a comparison between summer and winter or, provided that conditions change, it could be done over a week. At its simplest, the intensity of the microclimate is the temperature difference between the area under examination and the surrounding control area.

Use digital thermometers linked to a data logger to read temperature. Use GPS to log locations and GIS software to map the temperature readings.

Establishing the existence of a heat island is statistically rather harder than might at first appear. It is not enough just to say that the mean temperature of the town is 20 °C and the mean temperature of the surrounding area is 19 °C. Such a small difference could be explained by chance. You must use a test of significance such as the Mann–Whitney $U$ test (page 147) to find whether any apparent temperature difference is statistically significant (in other words, has probably *not* occurred by chance).

**1** Microclimate of a wood: as with all projects, you will need to read up your subject in a specialist book first. The particular interest of woodland is the effect of different tree types, notably coniferous and deciduous, and the effect of season on microclimate. If conditions permit, you can also examine the effect of variations in size and density of trees on microclimate. Measure temperature, humidity (relative or preferably absolute), wind speed and precipitation in and out of the wood.

**2** Urban climate: Just establishing the presence of a heat island is one task, and you may be able to go on to establish the time of day (or time of year) at which the island's intensity is most noticeable. Different wind directions may affect the heat island. Of course, temperature is not the only variable affected by towns: humidity and wind characteristics are too.

**3** The microclimate of a hillside (where aspect, altitude and exposure will affect it) can be examined together with vegetation (phenology) – see also Project Suggestion 1, page 70. This requires a longer-term approach and probably therefore more automated instruments.

## 8.10 Climate data sources on the internet

The Meteorological Office website at www.metoffice.
gov.uk contains up-to-date weather charts for the UK
and satellite images for the UK and Europe. This site
contains historic data for 37 weather stations going
back to the nineteenth century.

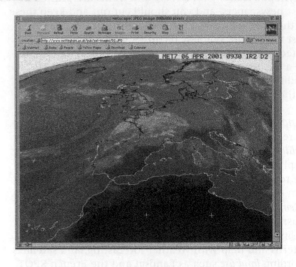

# 9 Primary data sources in human geography

## 9.1 Qualitative data

**Quantitative** research means collecting data which can be expressed in numbers – such as the number of residents in a village. **Qualitative** research means collecting data which cannot be easily converted into numbers, such as a study of people's opinions. Examples of qualitative research are:

1 Interviews, focus groups and oral histories (page 91).
2 Observations of people and places, recorded using field notes, film, photographs and sketches (page 23).
3 Textual analysis of printed newspapers (page 106), books, magazines and websites.

Qualitative data is just as valuable as quantitative data but harder to analyse. It has to be reduced to a few key conclusions supported by evidence from the interviews, observations and text analysis. This is why coding is useful.

### Coding qualitative data

Coding qualitative data means sorting it into categories as a way of helping analyse it. Imagine you have collected a large number of oral histories. You would read through the histories, jotting down themes as they strike you. When you have done this once you read it again, refining your list of themes (codes) and identifying those bits of text relevant to each theme (code). One theme could be the impact of *demographic change* on the area, for example. Another theme could be the way that *men and women* have different attitudes to an aspect of the past. These themes or significant points are 'codes' and you give each a name – *demographic change* and *gender* in this case. 'Codes', 'themes', 'categories' all mean something similar. Write down the codes and then tag (mark up) those parts of your oral histories which are relevant to each theme or code.

Approaching oral histories or interview data in this way – identifying codes as you inspect the data – helps to draw out significant themes from what at first sight may seem a fairly random pile of people's thoughts. Coding is simply a way of identifying main themes and grouping relevant bits of text under those thematic headings. Once you have done this, analysis of the text data becomes more straightforward.

## 9.2 Crowd sourcing

Crowd sourcing means using the internet to gather information from people – like an open questionnaire put onto Facebook or the many entries made on Wikipedia. Large sample sizes can be achieved at very little cost. However, the sample is biased – the people who respond to an internet data request are those who own a computer and have the time and enthusiasm to respond to such demands.

## 9.3 Questionnaires

Questionnaires are a set of pre-planned questions, to which the answers are written on a specially prepared form. They are the most widely used primary data source in human geography and are important for two reasons:
1 They enable us to find out information about people's opinions and behaviour that is not available from any other source (such as why someone moved house or how a decision was made to open a shop in a town).
2 They enable us to obtain information that is completely up to date.

The disadvantages of questionnaires are:

1 People do not like answering questions asked by a stranger.
2 Some people are particularly unlikely to be able to answer, such as mothers with young children on a shopping trip. Yet any survey that omits such groups is biased.
3 It has been shown that people do not tell the truth in surveys and opinion polls. They often give the answer they think you want to hear, or the answer that they believe sounds good.

*Writing the questionnaire*

Two examples of questionnaires are given in Figure 9.1, one good and the other bad. They illustrate some of the important points to remember when you write questionnaires.

1   Start by explaining the purpose of the questionnaire: most people are suspicious of them and reluctant to answer. The more convincing your explanation, the higher the response rate will be.
2   Plan the questionnaire so that the respondent is put at ease at the beginning. Leave more difficult or probing questions to the end.
3   Keep the questionnaire short. Use tick boxes where possible.
4   Know why you are asking each question and be careful not to omit anything you will need to know. It is impossible to go back later, find the same people and fill in the gaps.
5   Do not ask questions that are too personal, such as questions about age or income. It is normally possible to estimate age. You can form some estimate about income by asking questions about occupation, housing tenure and car ownership. People are usually reluctant to give their address, so ask for 'the area where you live' or for their postcode.
6   Do not ask questions that 'expect' a certain answer, such as question 8 in the bad questionnaire (Fig. 9.1).
7   Always issue a few (pilot) questionnaires to see how people respond. You will often want to make adjustments after this trial.
8   If you want to do statistical analysis on the results of your questionnaire, you must ask questions in such a form that the results can be expressed as numbers. This is one reason for giving people a limited choice in terms of their answers: 'Do you think this shopping centre is good, fair or not very good?' is a question that will enable you to express the answers as a percentage response to each of three categories. If you simply ask, 'What do you think of this shopping centre?' you may not be able to do this.
9   Do not include questions about information you can obtain from another source.

*Delivering the questionnaire*

1   Decide how you are going to administer the questionnaire.
a   Standing in the street and catching passers-by.
b   Going from house to house knocking on people's doors. This has the advantage that you are able to select the streets and houses you want.
c   Post the questionnaire into specific houses and pick them up later, or enclose a stamped addressed envelope. This method is quicker but more expensive and will not always achieve a very high

---

**A   Good Questionnaire**

Introduction: 'Excuse me, I am doing a school geography project. Could I ask you one or two quick questions about where you go shopping?'

1   How often do you come shopping in this town centre?
    More than once a week ❑ Weekly ❑ Occasionally ❑
2   How do you travel here?
    Walk ❑ Car ❑ Bus ❑ Train/Tube ❑
    Other_____
3   Roughly where do you live?_____
4   Why do you come here rather than any other shopping centre?
    Near to home ❑
    Near to work ❑
    More choice ❑
    Pleasant environment ❑
    Other_____
5   What sort of things do you normally buy here?
    Groceries ❑ Clothes/shoes ❑ Everything ❑
    Other_____
6   Do you ever shop anywhere else, and if so where?
    _____
7   Why do you go shopping there? _____
8   What do you buy there?_____
9   Sex: M ❑ F ❑ Age (estimate): under 20 ❑
    20–30 ❑ 30–60 ❑ Over 60 ❑
    'Thank you very much for your help.'

**B   Bad Questionnaire**

Introduction: 'Excuse me, but I wonder if I could ask you some questions?'

1   Where do you live?_____
2   How do you get here? _____
3   Do you come shopping here often? _____
4   Why do you come here? _____
5   Do you buy high- or low-order goods here?_____
6   Is this a good shopping centre and if so, why? ____
7   Where else do you go shopping? _____
8   Do you shop there because it is cheaper or nearer to your home? _____
9   How old are you? _____
    'Right, that's it then.'

**Figure 9.1 Two questionnaires, one good and one bad. Both were designed to find out people's shopping habits and were conducted in a town centre.**

response rate. A response rate of 30 per cent is typical for such a method. The other disadvantage with this method is that it does not allow the respondent to ask for clarification of a question. If you use a postal questionnaire, you must have a carefully written covering letter explaining the purpose of the questionnaire (Fig. 9.2).

2  Decide to whom you want to administer the questionnaire. You will often want to construct a stratified sample. For example, to ensure that all relevant sections of the community are represented, you may want to make sure that a proportion of the questionnaires are answered by men, a proportion by women, a proportion by younger people, a proportion by older, some by higher-income people, some by lower-income and so on. To find out what these proportions should be (for example, what proportion of the local population are of retirement age), you need to obtain census data (page 102).

3  Plan carefully when and where you are going to conduct the questionnaire. If, for example, you are hoping to interview people about their shopping habits, you must decide whether you wish to do a house-to-house survey or stop people in the High Street. You must decide which days of the week to do it and which times of day. Working people will not be around during the normal working day, so any questionnaire you undertake at this time is biased against them. The important point is to think these things through and to explain them when you write up your study.

   **Choose times and places that are safe.** It is best to work in pairs for what can sometimes be a rather unpleasant task. Some places will not permit you to conduct questionnaires on the premises. This is true of most covered shopping centres.

4  Decide how many questionnaires you are going to administer. If you have too few, you will not be able to draw any overall conclusions from the results.

```
School name
School address

                                    Date

Dear Householder,

I am carrying out a survey as an important part of
my A-level Geography project. The theme of the
project is the way in which the […] area of […] has
changed in the past twenty years. To find out about
this it is essential to obtain the views of local
residents.

The enclosed questionaire is short and, I hope,
interesting. I should be very grateful if you could
complete it and post it back to me in the enclosed
envelope. All the information you give will remain
anonymous and confidential.

I hope you can find time to help with this
important piece of work.

Yours faithfully,

                                    John Williams
```

**Figure 9.2  A sample of a letter to accompany a postal questionnaire**

## 9.4 Interviews, focus groups and oral histories

With a questionnaire the questions are fully planned beforehand, but often this is not the appropriate means of finding out information from people: we need a much more open-ended approach.

### Interviews

Most studies about the decision-making process, such as the reasons why a firm located where it did, or the thinking behind a local authority town plan, should involve an interview with the person who made the decisions or someone who can represent his or her views. Usually this will involve writing to the firm or local authority requesting an interview. Explain who you are and the purpose of the interview, indicate which days would be possible for you and how long it will take, and make clear what use will be made of the results.

Interviews should be approached in the same way as questionnaires – you go with a list of questions and paper to write the answers down (or a recording device – although some people dislike being recorded). The only difference is that you should listen carefully to the answers and be prepared to probe a bit. If you have arranged an interview with someone in advance, it is normal to write and thank them afterwards.

### Focus groups

Focus groups are groups of people who are got together

to discuss attitudes towards an issue. Political parties use focus groups to discover what the electorate think about issues. Focus groups should consist of a sensibly selected sample of representative people (i.e. a representative selection of different age groups, males and females, different ethnic groups, different levels of income), up to 12 people at a time (too many and it becomes harder to handle) and the discussion can be recorded or filmed. The researcher leads the focus group through the discussion using a series of open-ended questions – questions which allow a variety of responses.

*Oral histories*
Oral histories are accounts that people give of 'what life used to be like'. They are often recorded and are similar to questionnaires, except that they tend to be much more open-ended. The reason for this is that, although there will always be a set of specific questions you want to ask, there may be other important aspects of a person's life that had not occurred to you. This is why it can be profitable to start with a very open-ended question: 'What was life like when you were a teenager?'

Oral histories of this sort are especially useful in studies of demography (page 107) where we need to know about the influences of such things as age of marriage, number of children born to each parent, diet and health. Do not be afraid to use interviews and oral histories to find out about people's feelings – geography is not concerned solely with facts.

# 9.5 Perception studies

Traditional geographical theories assumed that people were rational, in full possession of all relevant facts, and made decisions in order to minimise effort or maximise economic advantage. In reality, people are not all-knowing and are influenced by considerations other than the need to maximise profit.

Studies of perception tend to fall into three categories:

1 how people perceive space;
2 how people evaluate their environment;
3 how people make decisions.

The best way to illustrate how these can be studied is by looking at what other researchers have done.

*How people perceive space*

**Mental maps**
The map of the world we hold in our minds is not the same as an Ordnance Survey map or an aerial photograph. One of the first detailed analyses of our mental maps was conducted by Kevin Lynch in cities in the USA. He simply asked people to draw a map of the city showing the places and features they knew (Fig.9.3). From these maps he was able to show that most people's mental map of a place was made up of five elements:

1 *paths:* routes, such as roads, along which people move;
2 *edges:* barriers around areas, such as woods or railway lines;
3 *nodes:* places where people cluster, such as market squares;
4 *landmarks:* distinctive features such as historic monuments;

5 *districts:* clearly identifiable residential neighbourhoods.

Lynch found that areas with plenty of all five elements were easy to get to know, whereas places without distinctive paths, edges, nodes, landmarks and districts were 'illegible' and it was harder for people to get their bearings.

Mental maps of this sort can also be used to show levels of knowledge of an area. Peter Orleans asked residents of three parts of Los Angeles to draw sketch maps of the city, bringing out the huge contrast between rich whites, black people, and poor immigrants from Latin America (Fig. 9.4).

The difficulty with studies of this kind is that it is hard to process the data, hard to combine many mental maps into one summary map, and impossible to do the sort of statistical analysis necessary to confirm the value of data. Nevertheless, mental maps are a very valuable method of showing both the way people structure space in their minds and how much individuals know about their home town. An interesting study would be to compare the mental maps of children of different ages, or of adults of different ages, or different income groups, or different ethnic groups. Mental maps are fun to do and thus easier to administer than questionnaires

**Subjective distance**
Donald Thompson did a questionnaire survey on a sample of 400 residents in San Francisco Bay. He asked them to estimate the distance in both miles and travel time to an attractive department store they shopped in regularly, to a less attractive discount store that was only used occasionally, and to a shopping centre they never shopped at (Table 9.1). He found that customers overestimated the distance to all shops, but the overestimates were greater in the case of the distances

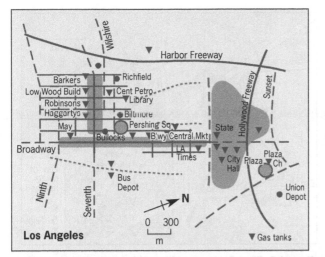

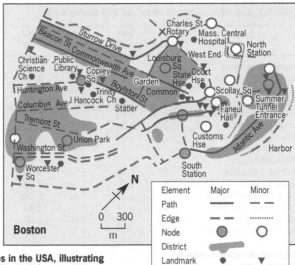

**Figure 9.3** Kevin Lynch's mental maps, drawn by residents of two cities in the USA, illustrating the presence of paths, edges, nodes, districts and landmarks in people's mental maps

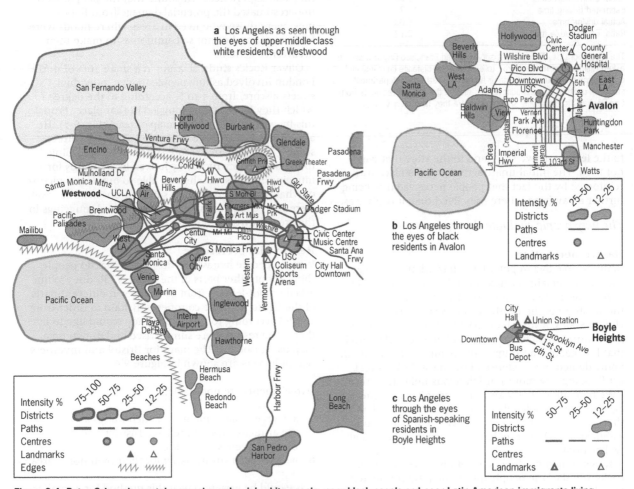

**Figure 9.4** Peter Orleans's mental maps, drawn by rich white people, poor black people and poor Latin American immigrants living in Los Angeles, illustrating different levels of knowledge about the city

**Table 9.1  Estimated v. actual distances and driving times**

|  | Albany | San Francisco | San Rafael | Walnut Creek |
|---|---|---|---|---|
| **Department store** | | | | |
| Estimated distance | 1.25 | 0.75 | 1.65 | 3.56 |
| Actual distance | 1.50 | 0.75 | 1.10 | 3.50 |
| Ratio: Estimated/actual | 0.83 | 1.00 | 1.49 | 1.02 |
| Estimated driving time | 5.50 | 3.45 | 6.07 | 8.42 |
| Actual driving time | 5.00 | 2.50 | 5.00 | 7.00 |
| Ratio: Estimated/actual | 1.10 | 1.38 | 1.21 | 1.20 |
| **Discount house** | | | | |
| Estimated distance | 1.68 | 1.46 | 1.67 | 3.65 |
| Actual distance | 1.50 | 1.25 | 1.10 | 3.50 |
| Ratio: Estimated/actual | 1.12 | 1.17 | 1.52 | 1.04 |
| Estimated driving time | 6.41 | 6.77 | 6.86 | 13.44 |
| Actual driving time | 5.00 | 3.50 | 5.00 | 7.00 |
| Ratio: Estimated/actual | 1.28 | 1.93 | 1.37 | 1.92 |
| **Regional shopping centre** | | | | |
| Estimated distance | | | 5.46 | |
| Actual distance | | | 3.80 | |
| Ratio: Estimated/actual | | | 1.44 | |
| Estimated driving time | | | 12.71 | |
| Actual driving time | | | 6.00 | |
| Ratio: Estimated/actual | | | 2.12 | |

The results of D. L. Thompson's survey in San Francisco Bay in which he asked residents of four suburbs to estimate the distance (in miles) and driving time (in minutes) from their home to an attractive department store, an unattractive discount house, and (in the case of residents from San Raphael) to a regional shopping centre they had never visited.

to the less attractive store and to the store that was not used. The initial unattractiveness of a place was reinforced by the fact that people perceived it as being further away from where they lived than it really was.

## The way people evaluate the environment

### Hazard perception

Many people live in potentially hazardous locations, such as floodplains or near the edge of unstable cliffs. R.W. Kates issued a questionnaire to residents along the seashore of the north-eastern seaboard of the United States, a stretch of coast that suffered damage from storms breaching the coastal defences. He found that few people faced up to the reality of the situation. Some denied the existence of the hazard ('We don't get flooding any more') or felt it was unlikely to recur ('We had one recently and so we won't get another for ages'). Some people felt the storms were cyclical and thus knowable ('They happen every ten years'), others assigned the whole problem to a higher power ('God's will be done').

Kates compared the results of his survey with the background of the people questioned. He considered how long a person had lived there, what their income was, how educated they were, and how many floods

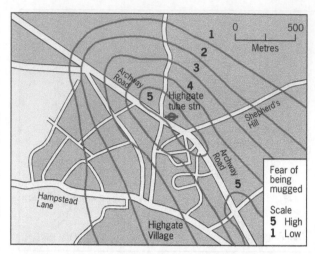

**Figure 9.5  Oliver Rock's map of fear of being mugged, drawn by residents of Highgate, north London**

they had experienced. He found that the people who underestimated the potential damage from floods most were those who lived in areas where floods were fairly common but not so common as to make their occurrence certain.

Oliver Rock's study of crime in a small area of north London involved asking people to give individual streets a score, from 1 to 5, according to the degree to which they feared being mugged in that place. From this he was able to draw a 'map of fear' (Fig. 9.5). He also showed residents a set of photographs of street scenes, asking them to score the pictures according to the degree to which they looked likely places for muggings. In this way he was able to build up a picture of what people perceived as safe and unsafe. The results were then compared with a map of actual muggings in the area, drawn from police data.

### Residential preference

Why do people live where they do? What do people think about other parts of the country? Rodney White asked school-leavers from 23 schools in different parts of the UK to rank the counties of Britain in order from 1 to 92 according to how much they would like to live there. The results were summarised and drawn on a map using isolines. The maps for Bristol and Inverness school-leavers are shown in Figure 9.6.

Two elements are noticeable:

a  a degree of agreement over such things as the attractiveness of the Lake District, the unattractiveness of Wales;
b  people had a strong liking for their own home area.

It would be easy to repeat this method for a smaller area, asking people to rank different neighbourhoods or streets. Even more interesting would be to ask them to rank areas according to particular attributes such as beauty, cleanliness, quietness, friendliness, accessibility.

A good alternative to putting places in rank order is to give them a score. In a research survey in Tanzania, university students evaluated parts of their country in terms of five characteristics, scaled from –3 to +3: cost of living, accessibility, environment, facilities, people (Table 9.2). These are called *bipolar semantic differential scales* (bipolar because the scale has two extreme ends, semantic differential because it is based on the difference between what the words mean).

## Knowledge of an area

Some studies have tried to show how imperfect people's knowledge is and how different groups of people have different levels of knowledge of the area where they live. Peter Orleans asked people from different ethnic groups to draw maps, marking on all the places they knew (Fig. 9.4). R.W.T. Chance gave people a list of settlements near their home town

**Table 9.2  Bipolar semantic differential scales used to measure perception of university students in Tanzania**

| | | |
|---|---|---|
| **Cost:** Expensive place to live in, saving from salary hard. | +3 +2 +1 0 –1 –2 –3 ⟷ | A very cheap place to live in, I can easily save from salary. |
| **Travel:** Very easy to get there, movement in or out no problem. | +3 +2 +1 0 –1 –2 –3 ⟷ | Very difficult to get to, and very hard to get in or out. |
| **Surroundings:** Very pleasant, an extremely nice place to live. | +3 +2 +1 0 –1 –2 –3 ⟷ | Very unpleasant surroundings, not a nice place to live in. |
| **Facilities:** Many things to do, spare time is no problem. | +3 +2 +1 0 –1 –2 –3 ⟷ | Not much to do in spare time, far from centre of things. |
| **Local people:** Friendly and very easy to get on with. | +3 +2 +1 0 –1 –2 –3 ⟷ | Very difficult for me to fit in and get on with people there. |

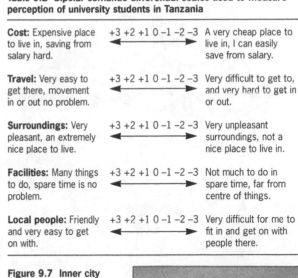

Figure 9.7  Inner city and suburb: a question of residential preference

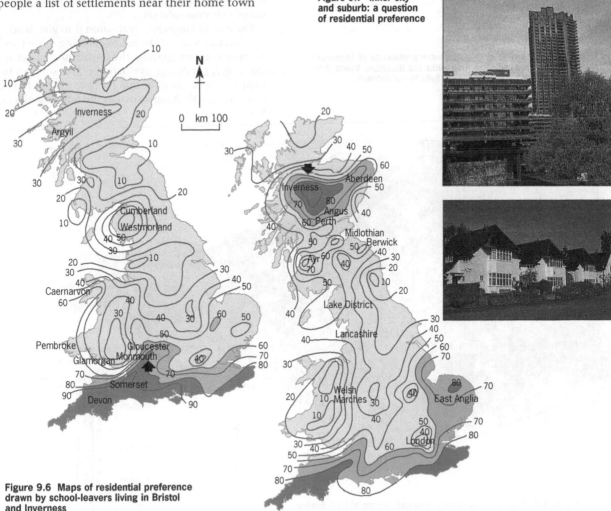

Figure 9.6  Maps of residential preference drawn by school-leavers living in Bristol and Inverness

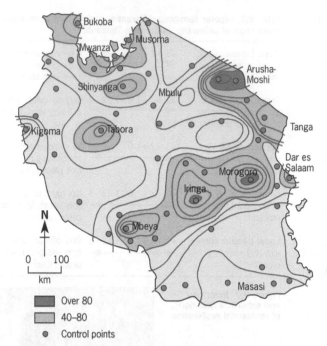

**Figure 9.8 A map of the residential preference of Tanzanian university students. The map picks out the major towns in the country, with rural areas being held in low esteem.**

and asked them to mark them on a base map. An alternative might be to ask people to list as many place names or street names as they can remember. In a project of this sort the interest lies in comparing the respondents' level of knowledge with their length of residence in the area, their age, income, education, and mobility. Such studies could also be part of a project on the use of recreational resources.

### Landscape evaluation

Perception analysis has been used by planners to evaluate landscapes. This helps to determine the level of development that can be permitted in the area. K.D.Fines applied the technique of landscape evaluation in East Sussex. Taking 20 photographs of a variety of views he asked 45 people to give a score (from 0 to 32) to each landscape shown according to its relative attractiveness. A photograph of a landscape with a score of 16 was issued as a guide. Fines then travelled around East Sussex giving areas a landscape score on the basis of the scores given to the twenty sample photographs (Fig. 9.9).

The aim of landscape evaluation is to give landscapes an objective score. This makes planning more precise because you can quantify the impact of changes to a landscape. Landscape evaluation of this sort can be used in A-level projects that try to assess the impact of environmental change, such as the building of a new road.

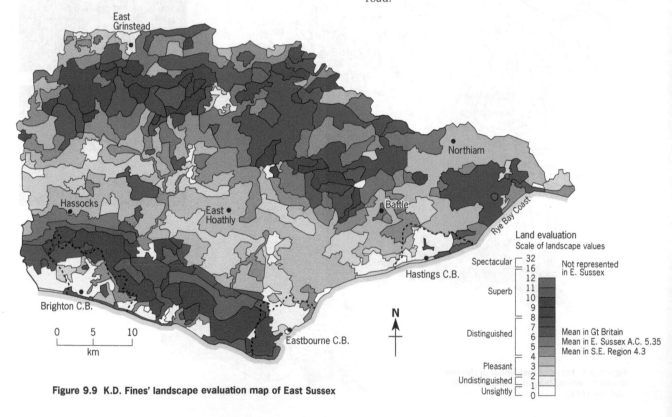

**Figure 9.9 K.D. Fines' landscape evaluation map of East Sussex**

**1** The purpose of this form is to measure your spontaneous responses to this view by means of adjective-pairs.

**2** By circling a number between 1 and 7 for each adjective-pair scale, indicate how strongly YOU feel these words describe the view.

**3** Please go down the list in order, doing each adjective-pair as it is encountered; do not leave any blank.

| | | | | | | | | |
|---|---|---|---|---|---|---|---|---|
| WET | 1 | 2 | 3 | 4 | 5 | 6 | 7 | DRY |
| UNEMOTIONAL | 1 | 2 | 3 | 4 | 5 | 6 | 7 | EMOTIONAL |
| UGLY | 1 | 2 | 3 | 4 | 5 | 6 | 7 | BEAUTIFUL |
| INTERESTING | 1 | 2 | 3 | 4 | 5 | 6 | 7 | BORING |
| BRIGHT | 1 | 2 | 3 | 4 | 5 | 6 | 7 | DULL |
| OBVIOUS | 1 | 2 | 3 | 4 | 5 | 6 | 7 | MYSTERIOUS |
| HARMONY | 1 | 2 | 3 | 4 | 5 | 6 | 7 | DISCORD |
| COLD | 1 | 2 | 3 | 4 | 5 | 6 | 7 | WARM |
| SOFT | 1 | 2 | 3 | 4 | 5 | 6 | 7 | HARD |
| FRUSTRATING | 1 | 2 | 3 | 4 | 5 | 6 | 7 | SATISFYING |
| PRIVATE | 1 | 2 | 3 | 4 | 5 | 6 | 7 | PUBLIC |
| STATIC | 1 | 2 | 3 | 4 | 5 | 6 | 7 | DYNAMIC |
| DISLIKE | 1 | 2 | 3 | 4 | 5 | 6 | 7 | LIKE |
| UNSTIMULATING | 1 | 2 | 3 | 4 | 5 | 6 | 7 | STIMULATING |
| FULL | 1 | 2 | 3 | 4 | 5 | 6 | 7 | EMPTY |
| PLEASANT | 1 | 2 | 3 | 4 | 5 | 6 | 7 | UNPLEASANT |
| WEAK | 1 | 2 | 3 | 4 | 5 | 6 | 7 | STRONG |
| DISRUPTIVE | 1 | 2 | 3 | 4 | 5 | 6 | 7 | PEACEFUL |
| COLOURFUL | 1 | 2 | 3 | 4 | 5 | 6 | 7 | COLOURLESS |
| DISORDERED | 1 | 2 | 3 | 4 | 5 | 6 | 7 | ORDERED |
| SIMPLE | 1 | 2 | 3 | 4 | 5 | 6 | 7 | COMPLEX |

**Figure 9.10 Scale for quantifying feelings about the landscape**

Another way of finding out how people feel about a landscape is to use a bipolar semantic differential scale. Figure 9.10 is a form used to test reactions to a natural stretch of river compared with a stretch straightened and dredged as part of a flood control scheme.

### How people make decisions

The way in which decisions are made, such as how a farmer decides what crops to plant, how a retailer decides to open a shop in a particular town, why a local authority decides to build a road, are usually studied by means of questionnaires and interviews. On page 130 is a description of a study that looked at the goals and values of farmers.

Allan Pred produced a useful diagram, called a decision-making matrix, which suggests that the success of a locational decision in economic terms depends on two things: the quantity and quality of information available to the decision-maker, and his or her ability to use this information (Fig. 9.11).

Perception studies are, like questionnaires and interviews, useful ways of getting closer to the reasons why people behave as they do. As with questionnaires, a poorly worded set of tasks will produce an unreliable set of answers. There is the danger that differences between people's answers may be due less to their differing perceptions of the place and more to differing perceptions of the meaning of the question.

After completing the questionnaire for a number of people, the summary form below can be used to calculate the overall 'attractiveness' score of the river and flood scheme.

**QUESTIONNAIRE SUMMARY**

| Scale values | Number of responses 1 | 2 | 3 | 4 | 5 | 6 | 7 | Mean value(m) | Weighting factor | Score |
|---|---|---|---|---|---|---|---|---|---|---|
| Unemotional/Emotional | | | | | | | | | 8–m | |
| Ugly/Beautiful | | | | | | | | | 5(m) | |
| Obvious/Mysterious | | | | | | | | | 3(m) | |
| Harmony/Discord | | | | | | | | | 5(8–m) | |
| Cold/Warm | | | | | | | | | 4(m) | |
| Soft/Hard | | | | | | | | | 3(8–m) | |
| Frustrating/Satisfying | | | | | | | | | 5(m) | |
| Private/Public | | | | | | | | | 2(8–m) | |
| Dislike/Like | | | | | | | | | 5(m) | |
| Unstimulating/Stimulating | | | | | | | | | 3(m) | |
| Full/Empty | | | | | | | | | 8–m | |
| Pleasant/Unpleasant | | | | | | | | | 5(8–m) | |
| Disruptive/Peaceful | | | | | | | | | 4(m) | |
| Disordered/Ordered | | | | | | | | | 3(m) | |

**Total score:**

After you have calculated the mean value of the responses, multiply each by the corresponding weighting factor to obtain the score. For example, in row 1, subtract the mean from 8; in row 2, multiply by 5.

View scores are obtained by adding together the weighted means, which can range from 49 (least attractive) to a maximum of 343 points. The total score is divided by 3.43 to give a value relative to a maximum of 100: e.g. 220.9 ÷ 3.43 = 64.4. The closer to 100, the more attractive the river.

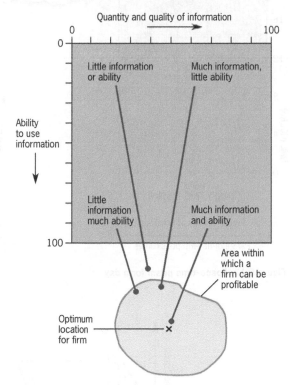

**Figure 9.11 Pred's decision-making matrix. The information about the different possible locations, as well as the ability to use this information correctly, determines how successful the chosen location turns out to be.**

## 9.6 Space–time diaries

You can do interesting projects to compare the ways in which different groups of people spend their time. How long do different people spend eating, working and sleeping, for example? How does the greater potential mobility of a well-off person affect the distances they travel in a typical day compared, say, with a low-income mother with young children?

To do such a project, first decide which groups you are going to compare. If you are interested in the impact of income, then it is important that the people you are comparing should be of a similar age and ethnicity, living in rather similar parts of a settlement, so that the only aspect that is different is income. Census data can help here. On the other hand, if you are interested in the impact of age, then all the people

you look at should be of a similar income. Ask your respondents to keep a space–time diary, listing what they did at different times of day, and where they went (Fig. 9.12). A week of such a diary is probably as much as most people are prepared to do. Draw maps to show the movement of people. Draw a space–time prism (Fig. 9.13) or a space–time path (Fig. 9.14), a three-dimensional graph with distance on the horizontal axis and time on the vertical axis.

You can use space–time diaries to form the basis of a recreational study (How do visitors move about a site? How long do they stay?) or at a retail park or shopping centre (Do customers shop at only one store, or move from one to another?).

**Figure 9.12  Space–time diary**

| Time | What did you do? | Time began | Time ended | Location |
|------|------------------|------------|------------|----------|
| Midnight | | | | |
| 1 am | | | | |
| 2 am | | | | |
| 3 am | | | | |

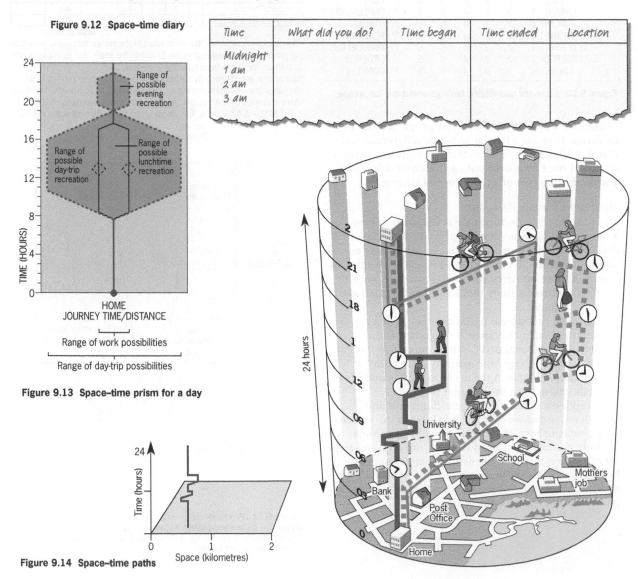

**Figure 9.13  Space–time prism for a day**

**Figure 9.14  Space–time paths**

## 9.7 Ethical considerations when collecting data about people

There are a growing concerns about the collection and use of data about people so be aware of the following:

**1** Informed consent

Before you collect data about people, if it might be of a sensitive nature you should have their informed consent – that is, the people knew why you were collecting the data and how it would be used.

There is a worry that the storage of huge amounts of personal information by Google, Facebook, Twitter etc (page 106) is an infringement of rights to privacy. This data can be sold to marketing companies who can then use your contact details and lifestyle choices to direct advertisements at you – even if you did not agree for the data to be used in this way! Many computer applications lack transparency about the way they will use the data you load on their sites. Just because an organisation can piece together a customer's life from their data trail does not mean it always should.

Another example of unauthorised use of data was the Dutch TomTom. The satellite navigation company aggregated data about their customer's car movements and speeds and sold it to local governments and infrastructure authorities. The Dutch police then obtained the data and used it to locate speed traps.

**2** No pressure on people to participate

People should never be pressurised to complete a questionnaire or take part in an interview, for example.

**3** Confidentiality

Most data collected about people is stored in a computer and can, therefore, be accessed by other people if you do not have in place technical safeguards to protect digital data from unauthorised access. People who collect sensitive data have to keep it secure.

Personal data should be recorded anonymously in order to protect individual's confidentiality.

**4** Care with vulnerable groups

These include children (do they really have the understanding necessary to give informed consent to be interviewed?) or old people.

# 10 Secondary data sources in human geography

## 10.1 Geographical Information Systems, geospatial mapping and GPS

A Geographical Information System (GIS) has a digital map as a base and layers of other digital information over it. Every item of information has a geographical co-ordinate, which enables it to be mapped and manipulated. This is a form of what is called geospatial mapping. The type of information held in Geographical Information Systems includes:

- census data: information about population and housing characteristics;
- postcode data: every group of 14 or so houses has its own unique postcode, and some data about housing and other characteristics can be linked to postcodes in the database;
- data collected by local authorities giving the location of such things as road repairs, derelict land, or the homes of people needing the services of social workers;
- data from the public utilities giving the location of drains, sewers, gas, water, telephone and electricity lines;
- data collected by health authorities about illness and patterns of disease;
- Ordnance Survey maps, in digital form;
- data from remote sensing satellites, transmitted to earth in digital form.

Every piece of information (such as the location of youth hostels on an OS map) is given a geographical co-ordinate when it is entered into the computer. Storing GIS data on a computer in this way offers several advantages:

1 Data can be easily transferred from one computer to another.
2 Maps and overlays can be easily updated. It is time consuming to record the location of new telephone cables on existing paper maps; it is far quicker to do this on a computer screen.
3 Data can be selectively mapped. Having input all the information from an Ordnance Survey map, for example, it would be possible to print out just the location of streams.

4 Data can be correlated quickly. It would be possible, for example, to compare the location of the homes of children admitted to hospital with asthma with the distribution of main roads, to see whether exhaust pollution could be a factor.

GISs on computers, tablets and smartphones has become commonplace as a result of three interrelated developments:

1 A huge increase in the amount of data being generated by governments, local authorities, businesses and others. A good example of this is satellite remote sensing data, which arrives in a constant stream day and night (page 102).
2 The development of computers powerful enough to handle the volume of data, including software that enables the computer to draw maps – computer-assisted cartography.
3 The increasing need for rapid availability of selected data. It is estimated that 80 per cent of the data stored by local authorities, for example, is locational in some way, and is crucial for planning decisions of all kinds.

The website www.magic.defra.gov.uk is a good source of free GIS on the internet: you will see many layers of environmental information about all parts of the UK.

Crowd sourcing (page 89) has been used to create maps. One example is OpenStreetMap which encourages people to download local geospatial data onto a world base map using local knowledge, GPS devices and aerial imagery.

Global Positioning Systems (GPS) are used for navigating, especially in cars (Sat Nav) but hand-held GPS receivers can also be used for mapping. A GPS receiver collects data from at least four satellites and this gives the map co-ordinates of the receiver, normally to within 10 meters. These can be recorded and used for making a map of the area being studied – the data you collect at any place can be given a precise map co-ordinate. GPS data is used to create GIS data.

## 10.2 Maps

*Ordnance Survey maps*

For most fieldwork a necessary first step is to obtain Ordnance Survey maps of the area. These are sold through bookshops and the OS websites – www. ordnancesurvey.co.uk and www.ukmapcentre.com. They can be used as a sampling frame on which sample points are located, and are a valuable guide to land uses, such as the distribution of woodland, marsh and heath. Road patterns in towns help to identify types of housing area – a grid of straight roads with narrow streets often indicates an area of Victorian working-class housing, while crescent patterns and culs-de-sacs are more typical of interwar and postwar housing.

Ordnance Survey maps also enable you to study the overall pattern of settlements, as discussed on page 119.

The following types of map are available:

1:10 000 from www.ukmapcentre.com

1:25 000 Some areas have special tourist editions at this scale.

1:50 000

1:250 000

In addition, the OS can print out a map for you centred on any point of your choosing, and at a variety of scales. The OS Geographic Information Systems programme allows customers to request specially printed maps of specific variables whose location has been recorded on the computer, such as all the pubs in Newcastle, or all the land below 20 metres in Yorkshire.

Ordnance Survey maps have been published since 1801. You can buy old versions of their maps dating from 1840 and these can be viewed and purchased online from the website www.old-maps.co.uk.

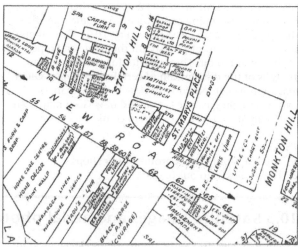

**Figure 10.2 Extract from a Goad map of Chippenham, Wiltshire**

*Land use maps*

The earliest twentieth century land use maps were based on ground surveys (Land Utilisation Surveys), the first in the 1930s which can be seen on the website www.visionofbritain.org.uk, and the second in the 1960s.

Recent land use maps are based on satellite images. The Centre for Ecology and Hydrology (www.cch. ac.uk) have land cover maps for the whole of the UK from 1990 to the present day (Fig.10.1). They are free of charge to students. Www.geoinformationgroup. co.uk also sell land use maps.

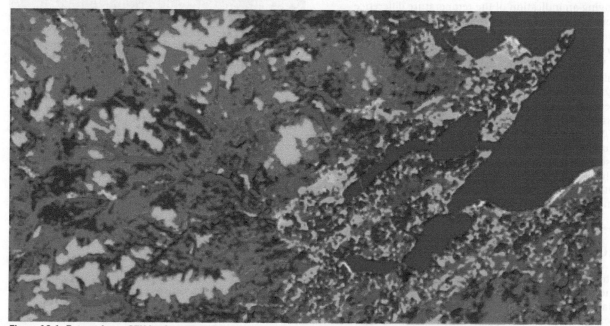

**Figure 10.1 Extract from CEH land cover map**

### Goad maps

Goad maps show the shopping centres of towns and cities with populations over 50 000 in Britain. They also publish shopping centre plans for some suburban shopping centres in large cities (Fig. 10.2). Goad maps show the location of shops with the name and type of shop marked on. This is useful for studies such as analysis of the distribution of certain shop types. Older maps can be compared with new to show changes within shopping centres. Old and new Goad maps can be obtained from www.experian.co.uk.

*The main drawbacks are:*
1 the maps only show ground-floor shops;
2 they quickly go out of date;

3 they do not always cover the whole area of shops in the centre of a town, especially if there are small clusters of shops separate from the main body;
4 the maps are expensive.

### Local authority maps

Local authorities may well hold other maps used for specific purposes, such as those that show conservation areas and planned developments in the area. Many local authorities make these maps available online.

### Town street plans

Street plans can be bought for most towns and cities. Apart from their obvious use in street surveys, they can act as base maps on which other features such as house type can be marked.

## 10.3 Satellite images and aerial photographs

Satellite images and aerial photographs can be used in field studies in two ways:
**a** For showing elements of the landscape that are not found on Ordnance Survey maps, such as crop distributions, areas of bare soil left after harvest, water pollution and private swimming pools.
**b** Landscape change can be studied by comparing older and more recent images.

The satellite does not record every detail of the surface, but looks at a block and gives a generalised colour or wavelength reading for that block; these blocks are called *pixels*.

Satellites send images at a variety of different wavelengths, including visible radiation (measuring the light reflected by different surfaces) and infra-red (giving an indication of the temperature difference between surfaces). The information is sent in *digital* format – a sequence of numbers that are then decoded to give the correct colour or tone of grey on the screen.

We cannot see infra-red radiation with our eyes so infra-red images have a *false colour* placed on them by the computer. In order to interpret the image it is necessary to find out how the false colours have been assigned.

You can use satellite images as the basis for individual studies, especially for examining agricultural land use. Bare soil appears clearly, as do certain crop types. Changes over seasons and over years can be measured. Other types of study for which you might use satellite imagery are recreational impact projects, water

**Figure 10.3 A Landsat false-colour photograph of Fife, Scotland. Pale blue and orange indicate grass pasture or cereal crops; pale yellow areas are upland grass and scrub. Exposed sandy soils show as pale green, while urban areas are dark blue – Dundee, for example, can be seen at the top, just right of centre.**

pollution studies, and soil–vegetation surveys.

Google Earth, the free satellite imagery package, provides an excellent way to present fieldwork data. Using Google Earth, it is possible to attach maps, photos, text and data to aerial photographs. Quikmaps.com shows you how to do this.

## 10.4 The National Census

The results from just one of the many questions in the census can form the basis of a good project. The National Census has been held in Britain every ten years since 1801, except 1941. The census is important to us because every household in the country is required to complete a census form. The census is therefore comprehensive, and for many purposes it is much the best data available. The topics covered by the census are those needed by central and local government for efficient management and for detecting social trends.

The most useful data is neighbourhood statistics which are accessed from www.neighbourhood.statistics.gov.uk. The most useful tables in the 2011 Census are those describing:

Population size
Population change
Population density
Age structure (the number of young children, etc.)
Ethnicity and religion
Country of birth
Marital status (e.g. married or cohabiting)
Household composition (e.g. the number of people who live alone, the number of families with young children)
Housing tenure (whether people own their homes or rent them)
House amenities – central heating
Type of house: detached, semi-detached, flat, etc.
Health
Car ownership
Method of travel to work
Distance travelled to work
Education and qualifications
Employment
Socio-economic class

## How to obtain census data

You may obtain 2001 and 2011 census data from the Office for National Statistics website.
Census data for the 1961, 1971, 1981, and 1991 censuses can be purchased from:
  Census Customer Services
  Office for National Statistics
  Titchfield
  Fareham
  Hants, PO15 5RR
  Telephone: 01329 444972
  E-mail: census.customerservices@ons.gsi.gov.uk

Some data from the pre-2001 censuses is available online.

## Stage 1 Decide on scale

The first stage is to decide which scale of areal units you are interested in - probably electoral wards, with populations averaging 5500 people. In the 2011 census the smallest areal unit was the Output Area (OA) which has about 300 people. In previous censuses these small areas were called enumeration districts.

## Stage 2 Obtain a boundary map

If you are looking for ward data, you should know the names of the wards you are interested in. Maps can be found on the ONS website and older maps purchased from Census Customer Services.

## Stage 3 Decide which types of data you need

Use the list to the left.

## Problems with census data

The main difficulties with using census data are as follows:

1  The census is taken only every ten years. By the end of any decade, the data is out of date.
2  Sometimes the boundaries of areal units (local authority districts, wards, etc.) change from one census to another. This makes it hard to compare data from one census to another.
3  Some questions we would like asked are not asked. For example, whereas the 2011, 2001 and 1991 censuses included a question on ethnic origin, the 1981 and 1971 censuses did not.
4  Some census tables are hard to understand without extra explanation. 'Born in the New Commonwealth' is an example of a phrase that is meaningless without a definition of 'New Commonwealth'.

## Project ideas based on census data

The detail and scope of the 2001 and 2011 censuses enable you to base a project very largely on this one data source. Because there are so many possible topics, it is important to start by deciding which aspects of your study area are of interest. For example, a coastal retirement area might be the place to study old people. An inner-city area could show interesting patterns of one-person and single-parent households.

There are two lines of investigation that are worth following:

1  Looking at the pattern produced by a census variable, such as the distribution of retired people.
2  Looking at the way a distribution has changed over time, such as the distribution of the places with the greatest increase in the number of retired people between 1981 and 2011.

In both cases you would be interested in the *reasons* for these patterns or changes and the *implications for planning policies*.

Worthwhile topics include:

Areas of population growth and decline
The causes of population change
Housing tenure
Patterns of single-parent families
Patterns of different ethnic groups
Patterns of socio-economic classes
Age structure
Limiting long-term illness
Unemployment and how it varies among different ethnic and age groups

## 10.5 Other central government population statistics

The Office for National Statistics also publishes statistics on a variety of topics such as birth rates and mortality, health and crime.

## 10.6 Parish registers

Parish registers, which record weddings (Fig. 10.4), baptisms and funerals in Anglican churches, have been kept in Britain since 1538. They are a readily accessible data source for the historical study of population. The information given in the registers is as follows:

*Baptisms*
Date of birth
Date of baptism
Name of child and parents
Address
Father's occupation

*Marriage*
Date of marriage
Name of marrying couple
Age of marrying couple
Occupations of marrying couple
Address of couple before marriage
Name of fathers
Occupation of fathers

*Burials*
Date of burial
Name of deceased
Age at death
Address at death

From this it is possible to study changes over time, of average age of marriage, birth rates, death rates, the distance between the premarital addresses of partners, the structure of occupations in a parish, and life expectancy. Parish registers are held by the local Anglican clergy, who will normally allow you to see them, or by the County Records Office. Some parish registers can be seen online and www.parishregister.co.uk can direct you. The main problems with parish registers as a data source are as follows:

1 It is not possible to tell the total parish population from them. This figure must be obtained from another source if birth and death rates are to be calculated.
2 Gaps often appear in the records, and many registers have been lost altogether.
3 Before the twentieth century the registers did not include dissenters from the Church of England, those who could not or would not pay the necessary fees, and, in the case of burials, suicides, criminals and the unbaptised. In the present century many people do not get married in church, preferring a registry office, many children are not baptised, and not everyone receives a Christian burial.
4 The boundaries of some parishes have changed over time.

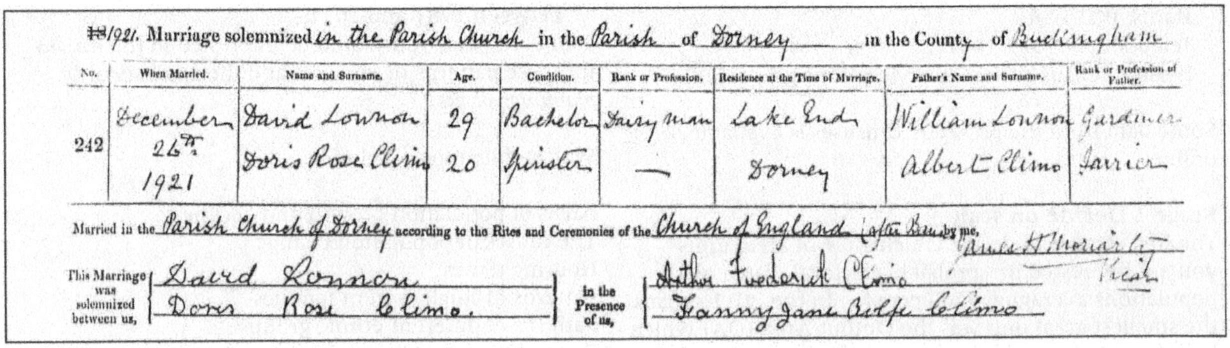
1921. Marriage solemnized in the Parish Church in the Parish of Dorney in the County of Buckingham

| No. | When Married. | Name and Surname. | Age. | Condition. | Rank or Profession. | Residence at the Time of Marriage. | Father's Name and Surname. | Rank or Profession of Father. |
|---|---|---|---|---|---|---|---|---|
| 242 | December 26th 1921 | David Lonnon | 29 | Bachelor | Dairy man | Lake End | William Lonnon | Gardener |
| | | Doris Rose Climo | 20 | Spinster | — | Dorney | Albert Climo | Farrier |

Married in the Parish Church of Dorney according to the Rites and Ceremonies of the Church of England after Banns by me,

This Marriage was solemnized between us, David Lonnon, Doris Rose Climo. in the Presence of us, Arthur Frederick Climo, Fanny Jane Rolfe Climo

Figure 10.4 A sample entry from a marriage register

## 10.7 Electoral registers

Electoral registers (Fig. 10.5) are compiled every year by local authorities. They list the name and address, by street, of all people eligible to vote. By comparing registers for two years it is possible to see how many people have moved house during this period. Ethnic groups with distinctive surnames can also be identified from the electoral register. The registers can be seen in local authority offices and main libraries.

The main limitations of electoral registers are:

1 A number of people do not register. This was particularly the case in the late 1980s and early 1990s, when people did not register in order to avoid having to pay local taxes.
2 Only those aged 18 or over are eligible to register. It is not, therefore, possible to use this source for calculating numbers living in each house.

```
           WREXHAM WEST CONSTITUENCY
                  POLLING DISTRICT
   NUMBER    NAME AND ADDRESS
                 001-ASH ROAD

     1       WALTERS GWEN
     2       WALTERS WILLIAM N.
     3       JOHNSON ANNE M.
     4       JOHNSON MALCOLM E.
     5       ANSELL ELSIE W.
     6       ANSELL JACK
     7       ANSELL JOHN
     8       SMEWIMG VIOLET E.
     9       STOKES ALISON
    10       STOKES PETER
    11       MILNE DONALD
    12       PALMER IRENE
    13       PALMER MERVYN H.
    14       EVANS GWILYM T.
    15       EVANS JOAN A.
    16       THOMAS DORIS G.
    17       THOMAS ROBERT J.
```

**Figure 10.5  Extract from an electoral register**

## 10.8 Rateable values and council tax bands

Every business has a rateable value, which is an index of the value of the property. It is based on the condition of the building, its location, and the value of the land it is on. These can be obtained from the Valuation Office Agency website.

Property value assessments for the local council tax (which came into operation in 1993) can be used in the same way. Councils will tell you the estimated value of every property you are interested in, or the value band the property falls into. These may also be found on the Valuation Office Agency website.

For a historical study, rate books can often be obtained for periods going back at least as far as the early nineteenth century. Rate books say who owned each property, who occupied it, what type of property it was, and its rateable value. They can be used to study property ownership and migration patterns, including rates of population turnover. Rate books are normally kept in County Records Offices.

The sale price of properties may be found on websites such as those of Zoopla and the Land Registry.

| | | **TRURO B.C. RATING DISTRICT** | | | | |
|---|---|---|---|---|---|---|
| ANALYSIS CODE | DESCRIPTION | ADDRESS (& NAME OF OCCUPIER, OR OWNER) | GROSS VALUE | RATEABLE VALUE | REFERENCE No. OF AMENDMENT |
| | | HERITAGE CLOSE | £ | £ | |
| 1050 | Flat | 1 | 330 | 250 | |
| " | do | 2 | 330 | 250 | |
| " | do | 3 | 330 | 250 | |
| " | do | 4 | 330 | 250 | |
| " | do | 5 | 330 | 250 | |
| " | do | 6 | 356 | 271 | |
| " | do | 7 | 330 | 250 | |

**Figure 10.6  Extract from a book of rateable values**

## 10.9 Directories

Kelly's Directories were published annually after 1845, for counties (divided by parishes) up to 1946 and for some larger towns up to the 1980s. There are two types of Kelly's Directories: trade directories, listing commercial premises including shops (Fig. 10.7), and residential directories, which list all residents in order down each side of every street. Trade directories for London (everywhere within the M25) were published up to 1991. Comparing directories over a number of years can form the basis of studies of retail change and population turnover. They are often kept by libraries and some may be accessed through kellysdirectories. com.

Kelly's trade directories usually group entries by parish and also give historical and population details of the parish. Parish boundary changes can make it hard to compare directories across many years.

For more recent years, you can use the Yellow Pages (www.yell.com) and Thomson Directories (www.

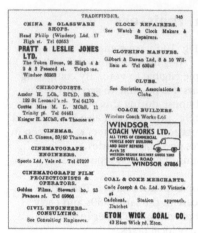

**Figure 10.7 Sample extract from Kelly's Directory**

thomsonlocal.com) for plotting the distribution of a particular industry, trade or service. If, for example, we wanted to study the changing location of insurance brokers in the City of London, Yellow Pages would be an obvious place to start.

## 10.10 Local newspapers

Newspapers can be used as an information source for population and settlement studies in a number of ways:

1 They provide information about the locations of local events and local businesses (through advertisements), giving an indication of the sphere of influence (page 120) of the town concerned.
2 Most local newspapers have a large section devoted to advertisements put in by estate agents. By noting house prices and the types of property advertised in different areas of a town, you can build up further information about the local private housing sector.

3 All local newspapers provide coverage of the important planning decisions reached by local authorities and analysis of the different views and interest groups involved in controversial issues such as road building and office schemes. These may well suggest possible topics for study, as well as being a source of information in themselves. Where libraries have kept back copies of the local papers, or extracts from them, these can be a good source of historical information especially when available to search online.

## 10.11 Planning data

Local authorities are an obvious starting point for data about the local environment. All have plans that control land uses in the area.

In addition to these, you will find planning studies that provide material on environmental assessments, sustainability, flooding, the local economy, retail and transport. Separate local authority departments hold information on housing, leisure services, water supply, and education.

Most new towns built since 1946 in the UK were set up and run by Development Corporations based in the town concerned. They were responsible for acquiring the land, preparing land-use plans, and co-ordinating development in the early years of each town's growth. Similarly Urban Development Corporations have helped manage urban renewal in parts of Britain since 1981. Development Corporations published detailed plans, progress reports and information for the public.

## 10.12 Big data

'Big data' is a term which describes vast data sets stored digitally, for example the abundance of data measured by satellites orbiting the earth several times a day or the details of every shopping purchase made by people with a Tesco's card or an Amazon account, or details of

the sites that you have Googled. Such massive data sets can be used to establish correlations – such as the times of the day or week that certain types of people buy certain goods. Using data from the past, algorithms can predict the future.

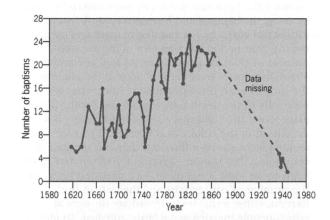

# 11 Urban and rural studies

This chapter illustrates the range of types of individual studies that can be undertaken under the general heading of human geography. The topics chosen are those that are commonly studied, that are often successful as project subjects, and that illustrate ways in which the data sources outlined in Chapters 9 and 10 can be used.

## 11.1 Demographic change

Demography is the study of population through the use of statistics. As a topic for an individual study, it has the merit that successful projects can be undertaken at various scales in most parts of the UK.

*The demographic transition model*
Studying changes in births and deaths is an interesting subject and can be very successful if the data are forthcoming. Such studies tend to rely on census data (Section 10.4) and, for periods before 1801, on parish registers (Section 10.6).

For studies involving parish registers, it is normally a good idea to take two or more adjacent parishes as your study area. The reason is that one parish may well have had too few baptisms, marriages or funerals in any one decade to produce an adequate sample size. Putting the data from several parishes together makes the sample size more convincing. For the same reason it is a good idea to pick large parishes rather than small ones.

It is also a good idea to structure the project in terms of a number of hypotheses. In the case of population change, these can be derived from the demographic transition model, which is a description of changes to the birth and death rates in England and Wales between 1700 and the present day (Fig. 11.1).

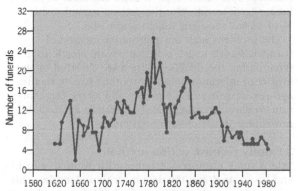

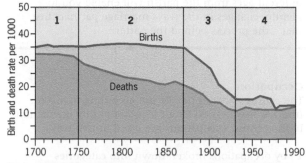

Figure 11.2 Graphs giving the number of baptisms and funerals conducted at the Church of St Andrew, Chedworth, Gloucestershire.

Figure 11.1 The demographic transition for England and Wales. The graph is divided into four stages: high death and birth rates; falling death rate (high birth rate, lower death rate); falling birth rate; low death and birth rates

As well as deciding which area to study, you will need to decide on the timescale of your study. Parish registers started in 1538, but you will probably want to study a more recent period, when the data are likely to be more complete and accessible. The period from 1750 to 1950 is the one that saw the demographic transition summarised in Figure 11.1. It is better to study a short period in depth than a long period superficially.

There are two approaches to the use of parish register data for demographic studies:

### 1 Aggregative method
This involves noting down the number of baptisms, marriages and burials each year. These are then plotted on a graph (Fig. 11.2). For each decade calculate the mean (such as the average age of death) and the standard deviation (page 143). For dates after 1801,

it is useful to find out the total population for your chosen parishes from the censuses (which started in this year). The census for 1851 gives the populations for all the censuses for 1801 up to 1851. If the population of a village is known, then you can estimate the birth and death rate per thousand people by using the baptism and funeral records. Dividing the number of births by the number of marriages over a twenty-year period gives an idea of the average number of children per mother. As well as changes in births and deaths over decades, look at the effect of the time of year: there might be seasonal changes, especially in the death rate. Historically, baptisms took place when the child was about a month old – ten months after the child was conceived. If you are looking for seasonal influences, then it is the date of conception that is more significant than the date of the baptism. In some areas there is a correlation between the number of conceptions and the state of the harvest. Other variables to record are the ages at which people married, gave birth, and died. To identify longer-term trends, plot the data on a graph using 5- or 11-year moving means (page 142).

The problem with the aggregative method of studying parish registers is that we are unable to tell to what extent the results obtained are distorted by the impact of migration in and out of the parish, by the age–sex structure of the parish, and by the fact that non-Anglican marriages and funerals in a parish do not appear in the church registers.

**2 Family reconstruction method**
Take a small number of families with distinctive names and just note down the details for those families. This method is very time-consuming, but helps avoid some of the pitfalls mentioned above. By tracing the history of one family's weddings, baptisms and funerals, you can study age at marriage, interval between marriage and birth of first child, interval between births, number of children per couple, and age at death.

For both the aggregative method and the family reconstruction method, it is important to try to explain changes in the number of births, marriages and deaths. Sometimes the registers themselves provide the

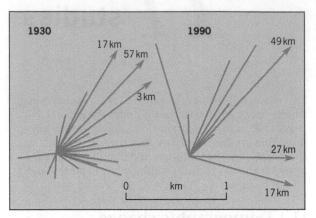

Figure 11.3 Ray diagrams with rays joining the premarital addresses of bridegrooms to the parish church of St Mark, Dalston, where they were married, 1930 and 1990

answers, because the clergyman made notes about events such as epidemics. More usually, you will need to consult local history society records or even textbooks that summarise national-scale events.

*Marriage distance*
Marriage distance is the distance between the premarital addresses of marriage partners. Find the marriage registers for your local parish church by consulting the clergy. For a period of, say, a hundred years, note down the age of both partners, their addresses before they got married, and the occupations of their parents. Because marriages usually took place in the parish church of the bride, it is possible to draw maps showing the homes of bridegrooms relative to the parish church for which you have data (Fig. 11.3). From these, you can analyse marriage distance, the geographical orientation that marriage partners came from, and changes in both of these over time. It is also possible to relate marriage distance to social class (deduced from the job descriptions given in the register). Are lower-income men more likely to marry the 'girl next door'? Interviewing people of a range of different ages from the parish will enable you to identify changes in the ways marriage partners first met – the process behind the pattern.

# 11.2 Social structure of urban areas

The social structure of an urban area, the layout of different social classes, age groups and ethnic groups, can form the basis for a worthwhile study.

*Social class*
Patterns formed by different social-class groups in larger towns and cities are interesting in their own right and can form the basis of useful individual studies. The obvious starting point for such a study is census data.

**Occupation**
The census classifies people not according to income (there is no question about income) but by occupation. In the census, occupations are listed in two ways:

1 by occupations broken down into categories (called 'Industry') such as 'construction';
2 by socio-economic groups, including categories such as 'Higher Technical Occupations'.

## Housing tenure

Occupation is one way of gaining an idea of social-class distributions in a town. Another way is housing tenure. The three main forms of housing tenure listed in the census are:

1 **Owner-occupied.** These tend to be better-off householders. The census distinguishes between those who own their house outright and those who are still paying off a loan (mortgage) for it.
2 **Rented privately.** These tend to be young people and those who live in inner cities (which have a relatively large amount of privately rented housing available).
3 **Rented from the local authority,** or a Housing Association. These tend to be people on lower incomes.

Mapping housing tenure is likely to produce similar patterns to maps of social class, although the sale of many council houses since the 1980s reduced the potency of this as an indicator of class.

## Other variables

Other census variables that are useful as indicators of social class are:

Unemployment;
Limiting long-term illness;
Lacking or sharing use of bath/shower and/or inside WC;
No car;
No central heating.

If you are trying to identify the poorest parts of a city, no one of these census variables is ideal: it is best to take several indicators together (for example, map areas that have a high proportion of single parents, high unemployment and high limiting long-term illness).

Normally you will use ward-level data and choropleth mapping techniques (Fig. 11.8 and page 138). If you are studying a town with a population of 80 000 or more, patterns of social class can be compared with the well-known models of urban social structure such as those of Burgess, Hoyt, and Mann (Fig. 11.4). For cities with a population of more than half a million, it might be difficult to map the whole city, in which case you could take transects from the centre to the edge.

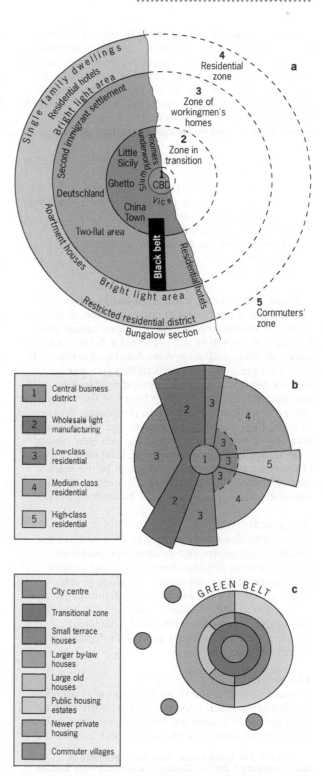

Figure 11.4 Models of the social structure of cities: (a) Burgess; (b) Hoyt; (c) Mann

After you have identified the existence of different social-class areas in a town or city, it can be interesting to take two adjacent but contrasting housing estates and measure the degree of interaction between them (Fig. 11.5). It is often the case that a council estate and a middle-class private estate next to it are close physically but distant socially. To measure this social distance, devise a questionnaire that asks residents about:

- whether or not they have any relatives living in the town and if so where;
- the addresses of their three best friends;
- the addresses of three people in the town that they had visited in the past fortnight;
- the school their children go to;
- the shops they use for general groceries;
- newspapers read;
- car ownership.

Studies about urban areas need some background information about their historical evolution. A project about the distribution of high- and low-income areas, for example, would be incomplete if it did not consider how and why some areas were developed for the rich, others for the poor – the original decision-making process. With all such studies you should look beyond the descriptive 'What?' and 'Where?' to the explanatory 'Why?' and 'How?', and then to the ethical 'What could or should be?' Start by speaking to your local librarian and finding the whereabouts of the best local history collection. Often this will be in the local history section of the main area library, which might be the main county library. Use old maps that, placed in historical sequence, show the evolution of the built-up area.

## Age groups

Census data also enable us to map the location of particular age groups and analyse aspects of their lifestyle. Recent censuses have a table that divides the population into 'life stages', breaking people down not only by age but also by whether or not they have children and whether or not the adults are living as a couple (married or unmarried). For larger settlements, mapping these variables will produce interesting patterns.

## The elderly

It would be interesting to study people of pensionable age. Using censuses, you can show how the number of older people in a town has changed over time. Look at the changing balance between men and women, the

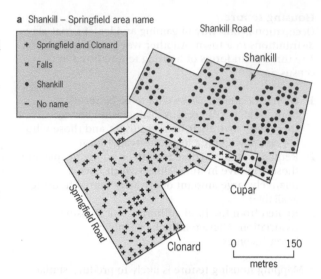

a  Shankill – Springfield area name

- **+** Springfield and Clonard
- **×** Falls
- **●** Shankill
- **–** No name

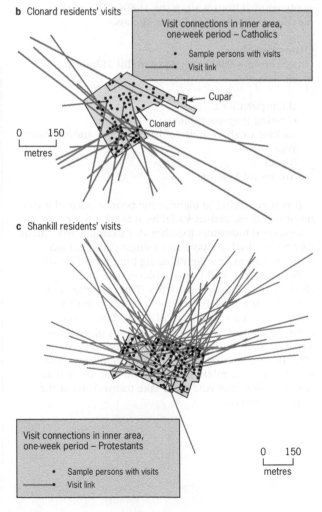

b  Clonard residents' visits

Visit connections in inner area,
one-week period – Catholics
- **●** Sample persons with visits
- **●—** Visit link

c  Shankill residents' visits

Visit connections in inner area,
one-week period – Protestants
- **●** Sample persons with visits
- **●—** Visit link

**Figure 11.5  F.W. Boal's research on the territoriality of neighbourhoods in Belfast. Clonard is a predominantly Catholic district, Shankill a Protestant neighbourhood, Cupar a street between the two. 11.5(a) shows the results of a survey asking residents if their neighbourhood had a name. 11.5(b) and (c) show lines joining a sample of residents with people whose homes they had visited in the previous week.**

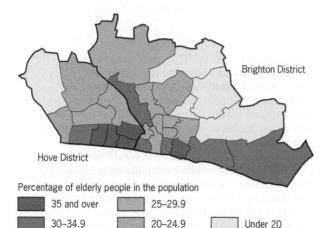

Percentage of elderly people in the population

■ 35 and over   ▨ 25–29.9

■ 30–34.9   ▨ 20–24.9   □ Under 20

**Figure 11.6 The proportion of elderly people in the wards of Brighton and Hove, 1991. There was a large variation within the wards, which is not evident from the map. (© Crown Copyright)**

number of widows and widowers. Map their location to show the degree of spatial concentration (Fig. 11.6). Look at their housing tenure, car ownership, the incidence of limiting long-term illness, and the degree to which they are still in employment. Using Yellow Pages, plot the location of old people's homes and day centres; to what extent do these correspond to the residential location of the elderly? What is the local planning authority's attitude towards the building of residential homes for the elderly? What were old people's homes used for previously? Use a questionnaire to find out the attitude of the elderly population to the facilities on offer to them.

### Young children
A similar and interesting project would be on the distribution of young children, relating this to social class and ethnicity, and seeing how it has changed over time. Try to identify the extent to which changes have been caused by nationwide shifts in the birth rate and to what extent the changes are due to selective migration into or out of an area. The highest number of young children is now to be found in the inner cities.

### *Ethnic groups*
Two aspects of many ethnic groups who live in British cities tend to make especially worthwhile studies:

1 The distribution of a group within a settlement, and the processes that lie behind these distributions.
2 The degree to which the group are or are not assimilating into British society. Assimilation means 'becoming like' the majority population and can be measured in a number of ways:
**a** residential segregation: the extent to which the group lives clustered together, apart from other groups;
**b** employment: is the group's employment profile like that of the population as a whole, or do its members specialise in certain trades or professions?

**c** language and religion: are these distinctive?
**d** dress: do they still wear ethnically distinct clothes? Sometimes there are differences in terms of assimilation between men and women and between one generation and another;
**e** diet;
**f** intermarriage: to what extent do people marry outside their ethnic group (so-called 'ethnic exogamy')?

The main data sources are as follows.

### 1 Census data
Before 1991 there was no census question about ethnicity. The only relevant question asked for the country of birth of the household head. Up to 1981 this question did in fact enable us to identify the majority of households of Asian, African or black-Caribbean ethnic origin: most had a household head who had been born abroad (these parts of the world are classified as 'New Commonwealth'). By 1991, however, many non-white heads of households were born in the UK. In the 1991, 2001 and 2011 censuses, therefore, there were questions about:
   The country of birth
   Year of arrival in the UK
   The ethnic group (such as white, black Caribbean, Indian, Pakistani, Chinese)
   Religion (2001, 2011)

Details are given about the following characteristics of ethnic groups: age structure; sex ratios; unemployment; those with a different address one year before the census; country of birth; socio-economic group and social class.

Ethnic groups should be mapped to bring out their location (Fig. 11.8) and the index of dissimilarity calculated to measure the degree of segregation.

It is also revealing to draw graphs and calculate correlation coefficients (page 144) to show the strength of the relationship between census variables (Fig. 11.9).

**Figure 11.7 Spitalfields, in London: home to immigrants of East European and Asian origin**

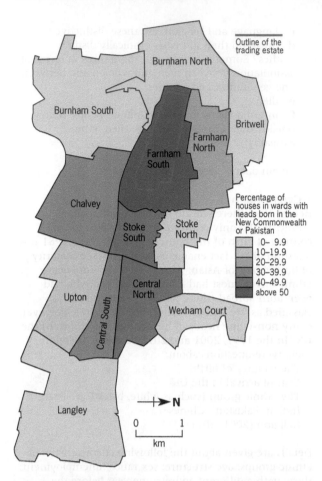

Figure 11.8 Percentage of households in wards of Slough, Berkshire, where the head of the household had been born in the New Commonwealth or Pakistan.

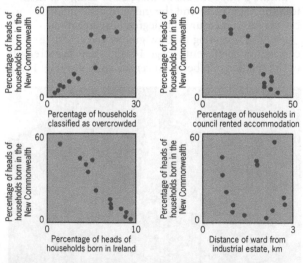

Figure 11.9 Graphs showing the relationship between pairs of census variables for wards of Slough, Berkshire. Each point is a ward.

THE ECOLOGICAL FALLACY There is a danger with correlations that is known as the 'ecological fallacy'. The ecological fallacy was first described by N.S. Robinson. Looking at the Census Bureau's nine regions of the United States, he found that there was a very strong correlation between per cent illiterate and per cent black: +0.946 (1.00 being a perfect positive correlation). From this finding it would be easy to conclude that many blacks were illiterate. However, when he looked at individual black people, he found that this was not the case: when colour and the ability of individuals to read were correlated, the correlation was a weak +0.203. The reason for the strong correlation at the regional scale was not that many black people were illiterate, but that many illiterate people of all ethnicities tended to live in regions that also had a high black population.

This, then, is an example of the need to take care when interpreting patterns. With all social geography projects it is important to ask *how* the observable patterns came about and what the *consequences* are of those patterns for society and for planners.

## 2 Electoral registers

Electoral registers (see Section 10.7) are useful for identifying ethnic groups with distinctive names, such as Asians and Jewish families. The significance of the electoral register is that it lists every adult in a street and thus gives us much more detail than the census. It is also published every year. Figure 11.10 has been drawn from electoral register data for Slough and shows both the rate at which Asians arrived in this road and the level of ethnic concentration within the street.

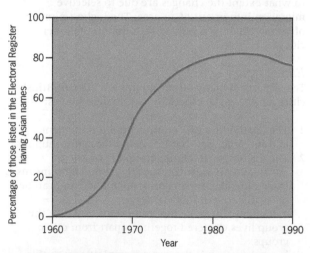

Figure 11.10 Graph drawn from electoral register data giving the percentage of names that appeared to be Asian, Woodside Avenue, Slough, 1960–90

### 3 Mapping ethnic institutions

Another way of establishing the presence of an ethnic group is to map distinctive institutions such as temples, mosques, synagogues, Asian shops and cinemas, and ethnic restaurants. A proportion of these might be identifiable from the Yellow Pages.

### 4 Questionnaires

Questionnaires are the only way of finding out certain crucial items of information such as: why a family came to the UK in the first place; why they are living in the particular district and house they are; experiences of racism; language used at home; religion; diet; attitudes towards marriage outside the group.

## 11.3 Inner cities

Because of their age and noticeable problems, inner cities are worthwhile study areas for projects. One approach is to see the extent to which Burgess's description of inner-city Chicago in the 1920s (an area he called the zone-in-transition) could be applied to inner cities today. Burgess described the zone-in-transition as being an area of:

poverty  
homelessness  
environmental deterioration  
first-generation immigrants  

crime and vice  
small-scale industry  
firms expanding in  
from the adjacent CBD  

To examine these characteristics:

**1** choose an older inner-city area that is not too big – say, one or two square kilometres.
**2** Obtain census data so that you can map variables such as population density, housing tenure, and the distribution of low-income groups and first-generation ethnic minorities. In order to show whether or not the study area has a particular character, you will need to compare it with other parts of the city. This can be done by using bar graphs (Fig. 11.11) or taking a transect out from the inner city to the suburbs (Fig. 11.12). Areas of poverty are found by mapping unemployment, low socio-economic groups and areas of rented housing. Census data can be used to show the extent to which particular ethnic groups have moved into or out of an area since 1961.

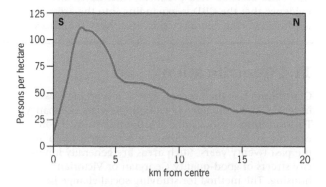

Figure 11.12 **Transect from the middle of London (Bank of England) to the Green Belt, population density, 2000**

**3** Electoral register data will enable you to plot the location of ethnic groups that have distinctive names.
**4** Estate agent valuations and council-tax bands can be used to map property values. Again, it is important to compare your study area with the city as a whole.
**5** Obtain old OS maps of the area in order to establish when particular streets were built. Where large areas have been rebuilt since 1950, either because they were bombed in the war or demolished for rebuilding (comprehensive redevelopment), old maps will show the previous street pattern. For London, Charles Booth's maps (Fig. 11.13) give the social class of streets at the end of the nineteenth century. Reproductions and reprints can still be purchased.

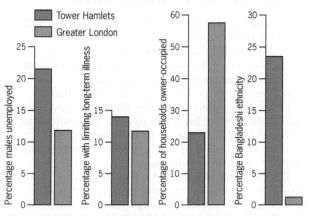

Figure 11.11 **Bar graphs showing the comparison between Tower Hamlets and London as a whole, 2001**

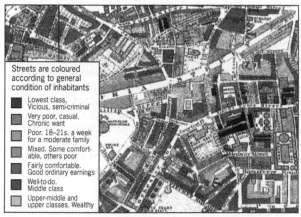

Figure 11.13 **A section of a Charles Booth map of 1889**

6 Obtain old trade directories, such as Kelly's, to show how the business function of the area has changed over the years.

7 Do a street survey. Work out the approximate age of buildings. Devise your own index of building quality, then map all buildings in the area. Plot the location of ethnic institutions such as mosques and temples. Record the existence of any manufacturing firms and describe their premises. Note down any evidence that the CBD is expanding into the area: are houses being demolished to make way for offices?

Record other land uses, such as street markets and wholesale markets, which might be there because the street is on the edge of the CBD.

8 Do a questionnaire to discover why people live in the area. If it seems that the area has attracted immigrants, why is this?

9 If there is manufacturing in the area, like the textiles quarter of Spitalfields in London, or Birmingham's jewellery district, speak to a sample of firms to find out the attractions of the area for businesses like theirs, as well as its disadvantages.

## 11.4 Gentrification

Gentrification is the process by which working-class inner-city areas change over to middle-class occupancy. This can form the basis of a worthwhile study if you can find an area that has undergone gentrification in the past twenty years. Such areas are generally inner-city streets of good-quality Georgian or Victorian housing. The method for studying social change is:

1 Obtain enumeration district level census data to map social change since 1960, looking at changes in housing tenure, socio-economic group, amenities, car ownership.

2 Use a questionnaire to discover why the middle classes moved into the area and what the long-established residents think of their arrival.

3 Conduct a street survey to map evidence of the modernisation of homes, such as new roofs, window frames, door furniture and extensions. Note the arrival in the area of 'middle-class' services such as wine bars and bookshops.

Social (demographic) change is discussed on page 107.

**Figure 11.14 A street of inner-city housing with gentrification in progress**

## 11.5 Council housing: architecture and indices of social breakdown

In the 1980s, Alice Coleman and the Design Disadvantagement Team from King's College, London, supervised a survey of 4000 blocks of council flats in order to test the hypothesis that there was a relationship between architecture and the behaviour of people who lived in the flats. It is possible to re-create some parts of this survey on a more limited scale, and those wishing to do so should read Alice Coleman's book *Utopia on Trial* (Hilary Shipman, 1985). The first task is to find a number of estates of contrasting character. **As long as these are safe to visit, and working in groups of at least two or three**, note down the following features:

1 The number of storeys: the higher the number, the worse the problems.

2 The number of flats that can be reached from each ground-level entrance: the higher this number, the worse the problems because, if many people have to use each entrance, it becomes impossible for residents to recognise strangers coming and going.

3 Whether the ground-floor entrance to the flats is visible from the street. Entrances that are tucked away and invisible are an invitation to crime and vandalism.

4 Whether or not there are locked doors at the entrance to the block of flats. Unlocked entrances allow strangers to move around easily.

5 The existence of overhead walkways. Alice Coleman found the existence of walkways (streets in the sky) to be among the worst of the design variables. Walkways allow muggers to move quickly out of sight and are strongly associated with vandalism and graffiti.

**6** Whether there are garages on stilts. These are areas where criminals can hide and are notorious for vandalism, litter and other environmental problems.
**7** The existence of public play areas. These are generally unsuccessful and associated with litter and vandalism.

Give each of these features a score: the higher the score, the more likely is the variable to create problems on the estate.

Next note down things that indicate a breakdown in standards of behaviour and maintenance on the estate, such as graffiti, vandalism, litter, urine smell at the bottom of stairwells.

For each of these devise a scale, say 0–5, according to the severity of the problem. Now correlate the design disadvantagement scores with the scores for indices of social breakdown, using graphs and correlation coefficients. Alice Coleman's results are shown in Figure 11.15.

You can use a questionnaire to find out the attitudes of residents on the estate to issues such as whether there is a sense of community, noise, security, crime, play areas for children, and the quality of estate management. Enumeration district level census data will give you details of social class, age structure, ethnicity, provision of household amenities, single-parent families, and car ownership. Arrange to visit the local authority housing department to find out about estate management and maintenance policies.

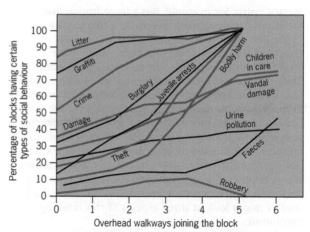

**Figure 11.15 The relationship between indices of social breakdown and the existence of overhead walkways that link blocks of flats to other blocks, from research conducted by Alice Coleman**

Alice Coleman interpreted the results of her survey to suggest that those elements of design that made spaces anonymous or hard to keep an eye on encouraged vandalism and crime. You might consider the extent to which architecture is to blame, or whether there are social problems in the area that contribute to the problem. How effective is estate management? What are the policy implications of your findings?

# 11.6 Retailing

Retailing studies divide into two types: those that study the distribution and character of the shops themselves, and those that study the shopping behaviour of customers. Many studies, of course, combine these two.

## The distribution and character of the shops

Historical studies, to show how the character of High Streets has changed, can be especially worthwhile. Choosing a reasonably large street of shops, use Kelly's Directories (page 106) to find the number and location of different types of shops at different periods in the past.

For more recent years the types and patterns of shops can be found by purchasing current and past copies of Goad maps (page 101). You can build hypotheses around some of the commonly observed trends in retailing of recent years.

**H1** *Independent retailers have been replaced by chain stores.* In the UK, the 10 largest firms now account for over a third of all retail sales. Because they have economies of scale and can negotiate such favourable prices with product manufacturers, they undercut the small independent shopkeepers.

**H2** For this reason *the average size of shops has grown*, with small fishmongers and grocers replaced by supermarkets.

**H3** *A proportion of shops has been replaced by quasi-retailing*: banks, building societies, estate agents, restaurants and fast-food outlets, printers and business services.

**H4** *Retail decentralisation is killing off town-centre shops.* One of the most important trends in retailing since the 1970s has been the construction of shopping centres outside town centres. Sometimes these are on greenfield sites on the edge of towns; more often they are on industrial or railway land that has become derelict within the town itself. The largest centres, such as Brent Cross in north London or the MetroCentre in Gateshead, are called *regional shopping centres* because they serve such a wide area. They benefit from the fact that sites away from overcrowded town centres have space to build car parks, lower land prices, and are often well connected to the regional road network. Because these are covered shopping centres they have a safe and secure environment – no traffic and no weather. They try to make a visit to the shopping centre an enjoyable 'leisure experience': the

MetroCentre, for example, has a large funfair and the shops are designed to resemble a theme park.

Smaller than these are *district centres,* such as Cameron Toll in Edinburgh, which usually have a food superstore as the anchor to attract other shops. They often contain *retail warehouses,* units with large floorspace selling durable goods, sometimes clustered in *retail parks* on former industrial estates. At the bottom of the hierarchy are the *freestanding superstores* that sell a wide range of products including food. Some retailers have chosen to keep their town-centre shops but develop subsidiary stores in out-of-town locations, whereas others have simply decided to build most of their new stores in out-of-town locations. Individual studies might look at three aspects of this retailing revolution.

a   What motivated those firms that have decentralised? Do the shops left in the town centre feel threatened?

b   How do customers view the relative merits of town-centre and out-of-centre shops? Is there an obvious difference between the attitudes and behaviour of middle-class car-owners and the relatively immobile lower-income or retirement-age population?

c   What strategy are the local planning authorities adopting? Are they taking measures to preserve the High Street? Are they preventing development of greenfield sites at the town edge? Or are they, by encouraging the development of retailing on derelict industrial sites, damaging the town centre?

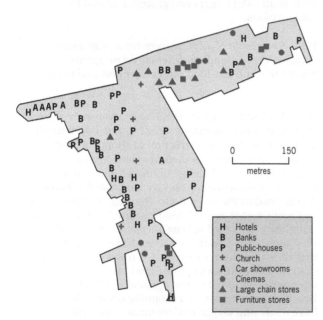

**Figure 11.16  The distribution of selected functions within central Cardiff**

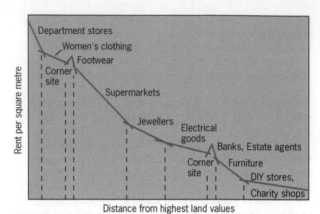

**Figure 11.17  A modified version of Scott's diagram showing the relationship between land value, High Street location and shop types**

**H5**   *Shops selling comparison goods (shoes, clothes, jewellery) tend to cluster in order to attract customers, while convenience goods shops (newsagents, grocers) disperse in order to reduce competition between each other* (Fig. 11.16). Apart from mapping the distribution of particular types of shops, you can give the distribution an index of concentration or dispersal by using the nearest neighbour index (page 152) or the index of dispersion (page 151).

One study successfully conducted a survey of shop managers, asking them to identify which of the following they would choose to locate next to because of the beneficial effect such proximity might have on trade: shop of a similar type, supermarket, building society, bank, restaurant, bus stop, car park, office, school, library. Managers of different types of shops were found to be interested in different types of locational linkage.

**H6**   *Land values are higher in the most accessible or central parts of the High Street, and shops locate in different parts of the High Street according to what they can afford* (Fig. 11.17). Near the centre, on the most expensive sites, are the highest-order shops such as department stores and supermarkets. Nearer the edge of the CBD we find lower-order shops such as charity shops and retail warehouses (which need a great deal of space) selling carpets, DIY materials, and furniture. Think about the shops that do not conform with this pattern as well as those that do.

Land values can be assessed by finding the rateable value of shops from the local authority and dividing by their floor area (page 105).

It would also be interesting in a study of this sort to measure pedestrian flow, counting the number of pedestrians passing specific points around the CBD, mapping these as a flow line (page 139). These flows can then be correlated with land values.

**H7** *Low-income areas have lower-order shops than high-income areas.* Here you might start by mapping certain census variables to identify high- and low-income areas, such as the proportion of the population renting property from the local authority. Then map the types of shops in each area, noting the sizes of shops, whether or not they are part of a national chain, and giving each shop a score (perhaps on a scale of 1–5) according to whether it looks scruffy or smart. It would also be interesting to devise a short shopping list of products that can be bought in any High Street in Britain, such as a bag of sugar, and to compare the price of this 'basket of goods' in the different High Streets. Are there differences between high- and low-income areas, or between high- and low-order shopping centres?

## Consumer behaviour

The ways in which people make decisions about which places to shop can be studied by using questionnaires, either house-to-house surveys or by standing in the High Street. Hypotheses that might be tested include:

**H1** *People travel to high-order centres to buy high-order goods but buy low-order goods nearer to home* (Fig. 11.18). Use the questionnaire to discover why people choose to shop where they do.

**H2** *Wealthier people and working people are more likely to ignore local convenience shops, preferring to buy all their shopping on a weekly multi-purpose trip to a large centre, travelling by car* (Table 11.1).

**H3** *People tend to overestimate the distance or time taken to travel to shopping centres that are unattractive and rarely visited, and tend to underestimate the distance and time taken to travel to shopping centres that are visited regularly* (page 93).

**H4** Compared with new out-of-centre superstores and retail parks, the traditional High Street can appear old, cramped, with poor car parking and an unpleasant

**Table 11.1  Relationship between consumer income and use of different levels of shopping hierarchy**

| Group 1 | | Group 2 | |
|---|---|---|---|
| Middleton (S. Leeds) | | Street Lane (N. Leeds) | |
| Low income | | High income | |
| 3 miles from CBD | | 3 miles from CBD | |
| Frequent public transport | | Frequent public transport | |
| 65 stores with 24 functions | | 78 stores with 25 functions | |

| | Destination | Group 1 No. | Group 2 No. |
|---|---|---|---|
| **Number of** | Hairdressing | 62 | 87 |
| **movements** | Shoe repairs | 61 | 90 |
| | Bank | 47 | 96 |
| | Cakes | 91 | 78 |
| | Cooked meat | 99 | 88 |
| **Direction of** | I *To local shopping parades* | % | % |
| **movements** | Groceries | 81 | 57 |
| | Meat | 75 | 54 |
| | II *To neighbourhood centre* | | |
| | Chemist | 92 | 78 |
| | Hardware | 50 | 23 |
| | Hairdressing | 52 | 40 |
| | III *To CBD* | | |
| | Convenience goods | 13 | 18 |
| | Groceries | 15 | 31 |
| | Cakes | 14 | 24 |
| | Adult clothing | 96 | 84 |
| | IV *To other retailing centres* | | |
| | Convenience | 6 | 25 |
| | Groceries | 4 | 31 |

environment. How do customers perceive the relative advantages of town-centre shops and out-of-town retail units?

**H5** Investigate how customers use out-of-centre retail parks and town-edge shopping centres. What is it that particularly attracted them to the centre? How often do they use the centre and how do they travel there? Once inside, do they visit just one shop or a range of shops? Plot the movement of people through a retail park or shopping mall, or ask shoppers which

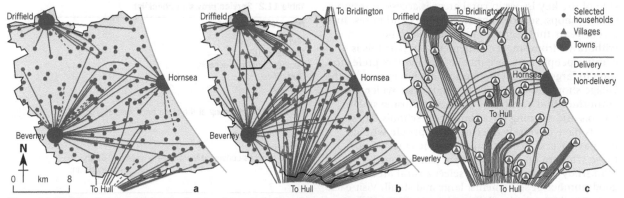

**Figure 11.18  Ray diagrams joining the homes of a sample of customers in Yorkshire to the places where they go to buy: (a) bread; (b) hardware; (c) shoes**

shops they had visited or intended to visit. Calculate the degree of linkage between stores: is it the case, for example, that one major food store draws people in, and they are then enticed to make casual purchases in other shops? Do customers compare shops that sell similar products, such as clothes shops?

**H6** How important are local convenience shops? Although there has been a trend towards shopping by car, there is evidence that local convenience shops are becoming increasingly important as well. Such shops sell mainly food, drink and other groceries, are self-service, and are open 16 or more hours a day, 7 days a week. They are growing in significance for three reasons:

**a** Many people only shop at the larger superstores once a week or so. They need a local shop to 'top up' with goods they have forgotten or suddenly discover they need.

**b** An increasing proportion of the population are single people living alone. They do not need much shopping and may therefore rely wholly on local convenience shops.

**c** An increasing proportion of women work, and they are likely to depend more on local shops that stay open outside working hours.

Many of these shops are run independently, but an increasing number are either partnerships between independent retailers and national wholesalers, such as Mace or Spar, or are linked to petrol stations. Use a questionnaire to discover the extent to which people use local convenience shops. What do they perceive their advantages and disadvantages to be? How do they get to the shop? How does the use of convenience shops relate to the customers' age, sex, housing tenure, whether or not they own a car, whether or not they work, and the distance from their home to the local shops (Fig. 11.19)?

## Questionnaire

How many times do you go shopping for food and groceries each week? _____

What proportion of these visits are to local shops and what proportion to larger shopping centres further away? Local _____ Further away _____

Which shops further away do you use? _____

What proportion of your weekly spending on food and groceries do you spend at local shops? _____

How do you get to your local shops? _____

What would you say are the main advantages and disadvantages of smaller local food and grocery shops compared with large supermarkets further away?

Advantages _____ Disadvantages _____

Street of residence_____

Age
☐ 16–24
☐ 25–34
☐ 35–44
☐ 45–54
☐ 55–64
☐ Over 64

Employment status
☐ Full-time job
☐ Part-time job
☐ Student
☐ Housewife
☐ Unemployed
☐ Retired

Sex ☐ Male ☐ Female

Number of children of pre-school age _____

Car availability for shopping
☐ Always ☐ Sometimes ☐ Never

**Figure 11.19 Questionnaire designed to examine the relative use of local convenience shops**

## 11.7 The impact of the closure of services in rural areas

One of the key issues in rural areas is access to services such as shops, schools, libraries, doctors and buses. In recent years these services have tended to be withdrawn from smaller settlements. In part this is because people are more mobile. Car owners prefer to shop in larger towns where prices are lower than in the village shop and the choice of goods is far wider. Local authorities find economies of scale by closing small schools and opening fewer but larger schools. These trends have adversely affected those people who do not have access to a car, especially the elderly and those who are on low incomes or unemployed.

To study service provision, select a rural area with a good number of settlements, large and small. Visit each and note down the level of service provision. Yellow Pages can also help with this task.

**Table 11.2 Service provision checklist**

**Name of settlement:**

Number of shops:
Number of GPs:
Number of primary schools:
Post office:
Chemist:
Bank:
Number of buses calling at the place each weekday:
Place of worship:
Public house:
Village hall:
Mobile services coming at least once a week: library:
food van:
other:

The survey will enable you to see how levels of service provision vary from place to place. In order to help to explain these patterns, obtain census data for the settlements. Then correlate service provision against settlement population size and the social-class structure of the settlements (Fig. 11.20). In order to use statistical analysis, you must devise an index that enables you to convert the results from Table 11.2 into numbers.

A questionnaire is now needed to discover the attitudes of residents to levels of service provision. Because this attitude is likely to vary according to whether or not they own a car, it is important to use a stratified sample, selecting both men and women, young and old, owner occupiers and council-house tenants. Find out where they go to obtain the services in question and how they get there. Discover what the main perceived problems are in terms of service provision. Are there differences between different social classes? Do women with children find themselves particularly immobile?

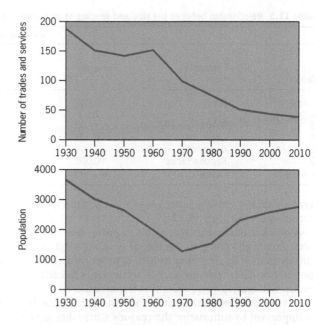

**Figure 11.20  Changes in service provision and population, 1930–2010, Tisbury, Wiltshire**

# 11.8 Settlement spacing and settlement hierarchies

Studies of the relative locations and sizes of settlements may take Christaller's central place theory as the starting point. The reason for this is that it is often best to tackle projects by testing one or several hypotheses – testing the various ideas in central place theory can form the basis of such hypotheses.

Christaller's theory is only concerned with the economic influences on the location of settlements that act as service centres. It has less relevance to settlements whose location was influenced by physical geography, mining, manufacturing or a host of other factors. Because it is impossible to find a large area where physical geography has had no influence on settlement location, and impossible to find an area where the only function of settlements is to act as service centres, we know from the start that we are never going to find a clear-cut example of a central place theory pattern of settlements (Fig. 11.21). Nevertheless, it would be helpful if we could choose an area to study that was similar to the sort of area Christaller looked at:

1   A relatively flat area with a large number of rural settlements, such as East Anglia.
2   A fairly large area, at least 50 square kilometres. This is important because we shall not find a settlement hierarchy (a range of settlements from small to large) in an area that is too small.

After you have chosen a suitable area, the hypotheses to test might be:

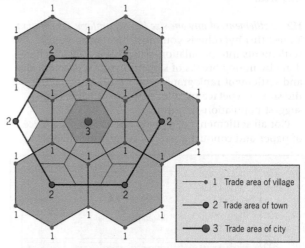

**Figure 11.21  Christaller's theoretical pattern of settlements**

**H1**   *Settlements form a hierarchy of size*, with many small settlements (low order) and few large (high order). The hierarchy found by Christaller in southern Germany is shown in Table 11.3.

To test this idea you need an OS map to identify the names of settlements in the area. Then you need settlement population size data from the census (page 102) or local authority planning departments. Write down the settlements in order of population size; this enables you to give every settlement a rank: the largest is rank 1, the second largest rank 2, and so on. Settlement size should be plotted against rank on

**Table 11.3 Relationship between the size and spacing of settlements in southern Germany according to Christaller**

| Order | Settlement type | Population | Size and spacing of settlements | | |
|---|---|---|---|---|---|
| | | | Number of settlements | Distance apart (km) | Tributary area size (km²) |
| 1 (lowest) | Hamlet | 800 | 486 | 7 | 45 |
| 2 | Village | 1,500 | 162 | 12 | 135 |
| 3 | Town | 3,500 | 54 | 21 | 400 |
| 4 | City | 9,000 | 18 | 36 | 1,200 |
| 5 | County city | 27,000 | 6 | 62 | 3,600 |
| 6 | Regional centre | 90,000 | 2 | 108 | 10,800 |
| 7 (highest) | Capital | 300,000 | 1 | 186 | 32,400 |

graph paper. Semi-log paper (page 133) is often used for this purpose (Fig. 11.22).

A quicker but cruder method is to group the settlements according to the size of typeface used on the Ordnance Survey map (which is proportional to population size). The number of settlements in each group will enable you to test this hypothesis.

Having established the pattern in your study area, it is important to summarise the reasons Christaller gave for expecting this hierarchy and the factors you believe are likely to influence settlement population size in this area.

**H2** *Settlements of any one size have a uniform distribution.* To test this hypothesis you must first group your settlements into population-size groups. This can be done by noting breaks of slope on your settlement size and settlement rank graph (Fig. 11.22) or by taking the size of typeface used for settlement names to suggest population-size groups.

Plot all settlements of a given size group on a piece of paper and conduct a nearest neighbour analysis

(page 152). This will tell you whether settlements are clustered, uniform or random in distribution.

Once again, you should find out why Christaller believed settlements would tend to have a uniform distribution, and comment on the influences on settlement location in your study area.

**H3** *Large settlements have high-order services as well as larger numbers of low-order services; small settlements have only low-order services.* Start by listing all the types of services likely to be found in a larger town. Select a sample of settlements, small, medium and large, and visit each, noting down the number of each type of service. For some services it might be even easier to look up in Yellow Pages the number in each of your chosen settlements. You might do a graph to show the relationship between settlement population size and the number of functions you have recorded. Calculate a correlation coefficient (page 144). Some settlements, such as towns not far from a large regional shopping centre, may have relatively few shops for their population. Others, such as coastal resorts, may have a small permanent population but many services.

In order to test the hypothesis, you must decide which functions are high order (requiring large numbers of potential customers every day) and which are low order.

**H4** *Settlements that act as service centres have catchment areas around them.* Large settlements have large catchment areas, small settlements have small. Settlements do not exist in a vacuum but exert an influence on the region around them. People from this region look to the settlement for work, entertainment, shops, hospitals and other services; the settlements look to the region for labour, customers and recreation. This is called the 'catchment area' or 'sphere of influence'.

To discover the catchment area of a settlement, you can explore many ideas including:

- Stand in the main shopping street, or outside a major peripheral superstore, of a settlement and ask customers where they live.

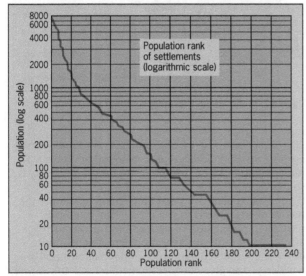

**Figure 11.22 Graph showing the settlement size relationships for settlements in south-west Wisconsin, USA**

- Ask shops that deliver goods where they deliver to.
- For larger towns, map the routes of bus services that start or terminate in the town.
- If a town publishes a local newspaper, plot sources of news and advertisements in the paper.
- Ask larger local employers if they will give outline details of the catchment area for their workforce.

Although some of these methods may give you a clear line, others will merely result in dots on a map, as in the case of the addresses of customers, and you will need to draw a line around the outer limits of these points. The result will be a series of lines around the settlement. Higher-order services will produce wider spheres of influence than low-order. The catchment areas of adjacent settlements will often overlap. Figure 11.23 shows the methods used to plot the catchment area of Wrexham.

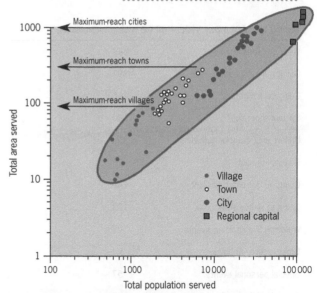

**Figure 11.24  Scatter graph showing the relationship between settlement size and catchment area, south-west Iowa, USA**

Figure 11.24 shows that larger settlements have larger catchment areas. However, the growth of out-of-town superstores and retail parks, often well outside towns or even near small settlements (at a major road junction, for example), is destroying the former relationship between settlement size and catchment area. In the past, local authority schools were often obliged to take their pupils from within a set catchment area – a line drawn on a map. In recent years government policy has been to encourage schools to compete with each other for pupils, as a result of which their catchment areas overlap more and more.

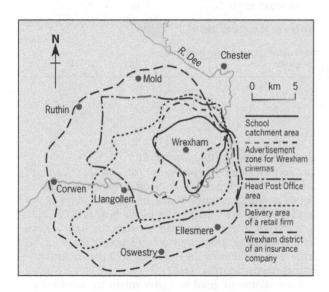

**Figure 11.23  Spheres of influence of Wrexham**

## 11.9 What is the impact of recreation and tourist pressure?

There has been a tremendous explosion in opportunities for leisure in the developed world since 1950. This has occurred because of a decrease in the length of the working week, an increase in the length of annual paid holidays, rises in real incomes, improved provision of leisure facilities such as golf courses and swimming pools, cheaper and quicker transport, and a fall in the cost of trips abroad because of the development of mass tourism and package holidays. The increase in car ownership is an essential aspect of this – from fewer than one-third of UK households in 1960 to over two-thirds by 2000. Over 20 per cent of UK households now have two or more cars.

Tables 11.4 and 11.5, and Figure 11.25, give a range of activities that might be the basis for a study of leisure.

Studies of recreation and tourism will normally consider two aspects of the subject:

**1** The fact that recreation and tourism are *concentrated in both time and space*. Ski resorts are under pressure in the winter but relatively empty in summer; beach resorts are crowded in summer, quiet in winter. Visitors to National Parks tend to concentrate in relatively small parts, along well-defined trails. Why do the visitors do this? Where do they come from, and how do they use the area?

**2** This concentration in time and space implies that there will be *intense pressure on the popular areas*. What is the resulting impact on the environment, on local employment, on land values, and on culture?

**Table 11.4  2013 Day Visits – all GB residents (millions)**

| | |
|---|---|
| Visiting friends or family | 545 |
| Going out for a meal | 397 |
| Going on a night out | 256 |
| Special shopping | 190 |
| Undertaking outdoor activities | 256 |
| General day out | 218 |
| Going out for entertainment | 142 |
| Going to visitor attractions | 128 |
| Watching live sporting events | 93 |
| Other leisure/ hobbies | 98 |
| Special personal events | 57 |
| Special public events | 73 |
| Other day out for leisure | 68 |
| Taking part in sports | 65 |
| Day out to health/beauty spa | 17 |

**Table 11.5  Household expenditure on selected leisure items in the UK (2011)**

| Average weekly household expenditure (£) | |
|---|---|
| Alcoholic drink consumed away from home | 7.40 |
| Meals consumed out | 15.20 |
| Books, newspapers, magazines, etc. | 4.10 |
| Television, video and computer equipment | 3.90 |
| Package holidays | 18.20 |
| Cinema | 0.60 |
| Theatre, concerts | 1.10 |
| Subscription and admission to participant sports | 1.40 |
| Admissions to spectator sports | 0.60 |
| Horticultural goods, garden equipment, plants | 2.30 |

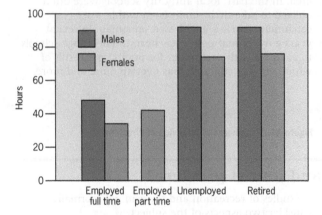

**Figure 11.25  Leisure time in a typical week: UK, 2011–12**

## Concentration in time and space

The concentration of visitors in time and space can be measured by doing a simple count at specific sample points at different times of day, on different days of the week, and in different months of the year (Fig. 11.26). In a town it might be possible to count the number of leisure activities, street by street. By giving these a score you could draw an isoline map of the intensity of these activities (Fig. 11.27).

A questionnaire used in a park might try to identify the following:

- Characteristics of the users:
  Age; sex; structure of the visiting group (alone, with family, with friends); job (which gives an indication of socio-economic status).
- Catchment area:
  Where the visitors have come from (are they people who have come from home, or non-locals on holiday?).
- Mode of travel.
- Motivation:
  Why they came (perceived attractions of the site); was this a single-purpose trip or did they do something else on the way there or back?

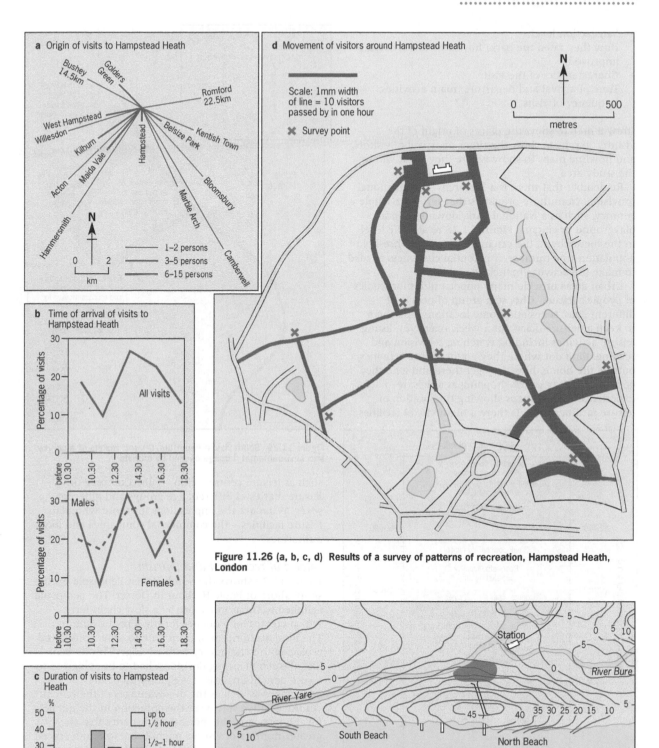

**a** Origin of visits to Hampstead Heath

Bushey 14.5km
Golders Green
Romford 22.5km
West Hampstead
Willesdon
Kilburn
Maida Vale
Acton
Hampstead
Belsize Park
Kentish Town
Bloomsbury
Marble Arch
Hammersmith
Camberwell

0   2
km

— 1–2 persons
— 3–5 persons
— 6–15 persons

**b** Time of arrival of visits to Hampstead Heath

Percentage of visits

30

20

10

0

before 10.30   10.30   12.30   14.30   16.30   18.30

All visits

Percentage of visits

30

20

10

0

before 10.30   10.30   12.30   14.30   16.30   18.30

Males

Females

**c** Duration of visits to Hampstead Heath

%

50
40
30
20
10
0

□ up to ½ hour
▨ ½–1 hour
▨ 1–2 hours
▨ over 2 hours

**d** Movement of visitors around Hampstead Heath

Scale: 1mm width of line = 10 visitors passed by in one hour

✖ Survey point

0   500
metres

N

**Figure 11.26 (a, b, c, d)  Results of a survey of patterns of recreation, Hampstead Heath, London**

Station
River Bure
River Yare
South Beach
North Beach

5   0   5   10

45   40   35 30 25 20 15   10

CTD  Defined by the intensity of tourist related activities and shown by means of a trend surface, i.e. a generalised surface similar to a contour line for height. The higher the figure the more tourist-related activities.

▨ CBD

Tourist retailing corridor linking CBD and CTD

N

0   1
km

**Figure 11.27  The central tourist district of Great Yarmouth, Norfolk**

- Satisfaction level:
  How they rated the park; how they feel it could be improved.
- Characteristics of the visit:
  Time of arrival and departure; main activities; frequency of visits.

Draw a map to show the places of origin of the visitors, graphs to show the times of arrival of visitors, and flowline maps to show where they went within the study area.

Remember that there is a hierarchy of recreational provision, from those providers that serve the whole country, such as a National Park, down to a local playground or cinema. Figure 11.28 relates the level in the hierarchy to the catchment area and threshold population (the number of potential customers needed to make the activity financially viable).

Urban areas provide many opportunities for studies of people's leisure. Choose a group of people of different ages, sexes and home locations. Ask them to keep a 'leisure diary' for a week (page 98), listing leisure activities including watching television and reading. Find out where they go for leisure activities outside the home, how they get there and who they go with. Do they count shopping as a 'leisure experience'? Draw maps showing the location of leisure facilities used. Is there a hierarchy of facilities

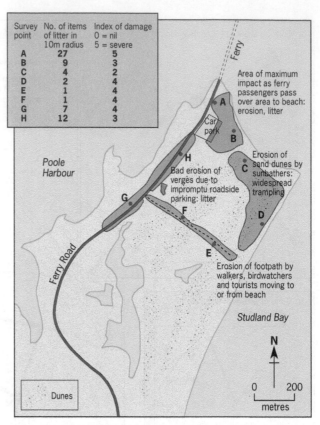

| Survey point | No. of items of litter in 10m radius | Index of damage 0 = nil 5 = severe |
|---|---|---|
| A | 27 | 5 |
| B | 9 | 3 |
| C | 4 | 2 |
| D | 2 | 4 |
| E | 1 | 4 |
| F | 1 | 4 |
| G | 7 | 4 |
| H | 12 | 3 |

**Figure 11.29  South Haven Peninsula, Dorset: results of a survey into environmental damage caused by tourists**

such as leisure centre – club – pub? Compare the leisure diaries of different age groups and different sexes. What are the implications for those who plan leisure facilities – the commercial companies and local authorities?

## Impact of recreation and tourism

Figure 11.29 shows the South Haven Peninsula on the south shore of Poole Harbour in Dorset. The peninsula is linked to the north shore by a slow chain ferry, which creates long queues for the ferry in summer. The local authority has, from time to time, considered replacing the chain ferry with a bridge. This decision, like all such planning decisions, had to be subjected to a cost–benefit analysis: a weighing-up of the advantages as well as the disadvantages of the scheme. On the positive side were the reduction in the queues for the ferry and the increased prosperity that the greater number of visitors might bring to South Haven and the Isle of Purbeck beyond. On the cost side was the possibility that removing the ferry queues would attract too many visitors, creating road congestion and ecological damage along the coast. Impact studies of this sort need to consider four elements: environmental impact, social and economic impact, perceptual capacity and compatibility.

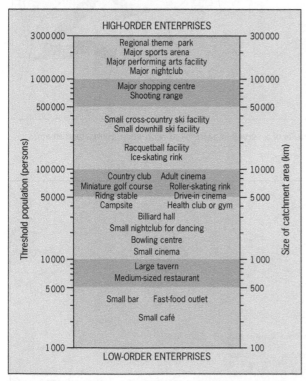

**Figure 11.28  Hierarchy of commercial recreation enterprises in the USA**

## Environmental impact

Because visitors tend to concentrate along specific paths in limited parts of recreational areas, the pressure on those areas damages the natural environment:

1 Trampling plants will kill off some species, reduce the abundance of others, and may increase those species that tolerate the effects of trampling. Figure 11.30 illustrates the likely pattern of a species as an environmental variable changes, such as temperature, soil moisture, or trampling by people. Plants have an upper and lower level of tolerance of such variables (Fig. 11.31). These levels of tolerance will differ from one plant species to another, and this changes species composition when people start walking over a previously undisturbed area.

2 Once plants have started to die back, soil erosion sets in. Figure 11.29 is an area of sand dunes. Erosion of the marram grass, which normally binds the sand with its dense network of roots, has allowed the wind to blow away sand, creating hollows called blowouts. The footpaths are eroded by walkers and compacted by trampling feet. The paths become filled with running water in heavy rain, which cuts gulleys in the footpath. Erosion of soil means that the humus-rich upper layers are lost, making it harder for plants to re-establish themselves.

3 As the conditions for plant growth deteriorate, so do the average height and density of plants.

4 Changes in the soils and vegetation will affect the variety and composition of animal and insect life.

These variables can be measured by a field survey:

1 Using a tape measure, record the width and depth of visible footpath erosion (Fig. 11.32). Note the evidence for blowouts and gulleying. Use a clinometer to measure slope angles (page 21).

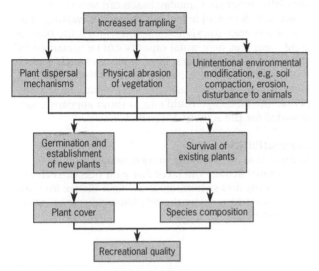

Figure 11.30 The impact of trampling

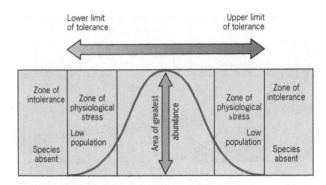

Figure 11.31 Species distribution curve

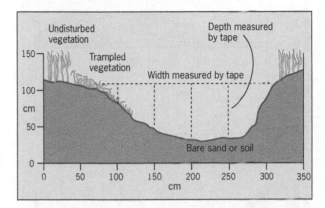

| Distance from centre of footpath | Depth of path (cm) | Depth of humus layer (cm) | Height of vegetation (cm) | No. of species in metre-square quadrat | Evidence of animal life |
|---|---|---|---|---|---|
| 0 | 71 | 0 | 0 | 0 | 0 |
| 50 | 53 | trace | 0 | 0 | 0 |
| 100 | 28 | 5 | 9 | 1 | 0 |
| 150 | 5 | 10 | 24 | 3 | burrow |
| 200 | 0 | 11 | 43 | 7 | burrow, droppings |
| 250 | 0 | 11 | 42 | 8 | insects |

Figure 11.32 Cross-section through a footpath: measures of erosion

2 Using a quadrat placed at different distances from the centre of the path, estimate the proportion of the ground that is bare soil, the number of plant species, the proportion of the ground covered by each species, and note any evidence of animal or insect life.

The results can be presented in a table (Fig. 11.32) or as line graphs (Fig. 11.33). The relative occurrence of plant species can also be presented as a kite diagram (Fig. 11.34). Kite diagrams are commonly used for showing transects of this sort: the data for any one plant type is plotted twice, once above and once below the horizontal axis, to give a visual impression of the relative rise and fall of species abundance.

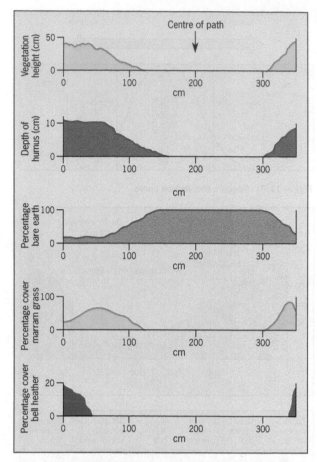

**Figure 11.33 Graphs showing the results of measurements taken across the footpath (see Figs 11.31, 11.32)**

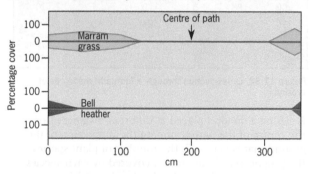

**Figure 11.34 Kite diagram showing the relative density of two plant types across a footpath**

Use correlation coefficients (page 144) to show the relationship between distance from the centre of the footpath and the other variables, such as number of plant species.

**3** Cut into the ground with a spade and pull back the soil to expose the soil horizons. Measure and draw these to compare the soil profile of eroded with uneroded parts.

**4** Obtain fifty or so wire nails 5–10 cm long and with a large head. On the top of each, solder 3 cm of fine wire. Insert the nails into the ground so that the head of the nail is flush with the ground surface and the wire is vertical. Place them at regular intervals across the footpath. Return after a week or so and record which of the projecting wires have been bent by walkers.

**5** Choose a number of representative survey points (A–H in Figure 11.29) and, at each site, count the number of pieces of litter within a certain radius. Record the degree of environmental damage, using a scale of your own.

**6** In a laboratory, grow a sample of different plant species in flower pots or trays. Devise a standard system that can act like the trampling of feet. Observe the relative ability of plants to withstand different levels of trampling. What impact does the trampling have on soil compaction? Does trampling prevent growth of new shoots? How long do the different species take to recover once the trampling stops?

### Social and economic impact

Construct a questionnaire for local residents to determine their attitudes towards tourists: who gains and who loses from tourism in the area? The local authority planning department should be contacted for an interview about its land management policies, as should the local tourist office (where there is one).

To estimate the impact of tourism on a local economy, compare a tourist resort with a non-tourist settlement of similar residential-population size. Compare the number of shops and other services in each, mapping the location of tourist shops, hotels and other tourist facilities.

### Perceptual capacity

All recreational sites have a limit to the number of people who can use them before the enjoyment of the area falls. Whereas a holiday beach can become quite crowded before people feel miserable, hill walking in a wilderness area quickly loses its attraction if the place is congested. This perceptual capacity can be measured by showing people photographs of landscapes such as a beach and a moor and, having specified the activity (sunbathing, walking), ask them to estimate how many people the landscape could take without appearing too crowded for the activity concerned.

### Compatibility

In areas that attract many tourists, some forms of recreational activity often conflict with others, such as water skiing and swimming. Draw up a matrix for your study area and consider the relationship between pairs of activities.

# 11.10 How do cities influence the rural areas around them?

This theme is a wonderful one because there are so many different possible topics and because such large parts of the UK are suitable as study areas – anywhere that is influenced by a large town or city. There are two possible approaches:

**1** Study the way the area has changed over time as the influence of the city has grown. This approach is *historical*.
**2** Examine the ways in which the characteristics of the rural area change as you move away from the city. For example, one might expect land prices to fall with distance from the city. This approach is *spatial*.

Among the many hypotheses you might test are the following:

## Agriculture
**H1** *Von Thünen's theory: land-use intensity falls away from the city*, with market gardening nearer to the city, arable and beef cattle further away.
**H2** *A high proportion of farms nearer the city have been bought by people who work in the city and use the farm as a nice home and a source of extra income.* These are known as *hobby farms*, and are often run by farm managers. They tend to concentrate on farming activities that do not need too much labour, such as arable crops or beef cattle.
**H3** *Farms near cities offer a high proportion of farm-gate sales and pick-your-own produce.*
**H4** *Farms near cities suffer from trespassers and the impact of urban developments such as motorways.*
**H5** *Farmland near cities has been turned over to other uses needing large amounts of space, such as golf courses and garden centres.* Government set-aside policies and other changes designed to reduce food production will encourage farmers to move away from agriculture.

These hypotheses can be tested by using land use maps (page 101) and questionnaires sent to farmers (it is often possible to work out their addresses from an OS map).

## Non-agricultural land uses
**H6** *The urban–rural fringe has been damaged by land uses that serve the population of the city*: waste-disposal sites (landfill), reservoirs, gravel pits, motorways, airports, shopping centres. The development of these land uses can be examined by looking at OS maps from various periods. You should visit local authority planning departments to discuss the planning approach to these necessary but disruptive land uses.

## Social change
**H7** *Settlements beyond the city have changed with the arrival of migrants from the city*. With improvements in road and rail networks, and higher levels of car ownership, people are increasingly able to live outside cities while commuting there for work. Since 1970 many jobs have also moved out of cities to the areas around. Rural areas have replaced the coast as the most popular destination for retired people. This movement of people from the cities to the rural areas is known as *counter-urbanisation*.

The original inhabitants of rural villages were farm workers and people who ran local services such as the shops and school. They tended to have low incomes, worked locally, had low levels of car ownership, and had many local friends and relatives. The newcomers from the city were often upper-middle class, they worked in the city, had high levels of car ownership and widely scattered friends and relatives.

The contrast between these groups can be studied by using census data, which will reveal population growth, changes in social class, age, housing tenure and the extent to which residents commute to work. The difficulty with a study of this kind is sorting out which changes have occurred because of the arrival of new residents and which have occurred because of rising standards of living among the original residents (for example, car ownership will have risen even without the arrival of the middle classes in the village).

Conduct a questionnaire to find out why the newcomers came to live there, how the different groups regard each other and how they differ in terms of lifestyle, mobility, job location, and the location of friends and relatives. Try to find out if the social changes have brought with them changes in the nature of local services, and the extent to which the housing stock has been upgraded.

Contact local authority planners and ask about the impact of planning controls on village growth. Some villages have been allowed to grow, and a field survey should be conducted to map the age of buildings. In others, building has been restricted, including those settlements that lie in the Green Belts.

A related theme, which can be a project in itself, is the impact of second-home ownership on rural villages. There are certain areas of Hampshire, Dorset, the Lake District, East Anglia and Wales, for example, where a high proportion of the houses in a village are second homes.

## Green Belts
Green Belts are areas of land around cities and larger towns within which there are controls on new building. The land is for the most part privately owned. Most of the Green Belts have been designated since the

Second World War, and today there are twenty of them in the UK. In most cases the aim of the Green Belts was to stop the city growing (preventing sprawl), either because it had simply become too big or because there was a danger of adjacent settlements merging.

Green Belts have been largely successful in preventing urban sprawl, though they suffer from a number of criticisms:

1   In some cases development that could not take place in the Green Belt has happened beyond the Green Belt instead: it has simply leaped over the

belt, putting pressure on the countryside.

2   By restricting urban expansion, Green Belts have contributed to the shortage of homes and high house prices in cities.

3   Green Belts are not positively managed, and large parts seem run-down or semi-derelict.

4   There are pressures to build in the Green Belt: the M25 around London was built through it, for example.

The methods discussed in this section might form the basis for a sound project based around an assessment of the strengths and weaknesses of an area of Green Belt.

## 11.11 The geography of fear and crime

The regular crime surveys carried out by the government have shown that many people have a fear of crime (Table 11.6). Very often this fear has a geographical component – some places are more frightening to people than others. This fact provides us with the possibility of mapping fear of crime and then assessing what factors contribute to this fear. To do this you will need to construct a questionnaire.

**Table 11.6  Fear of crime in the UK (*Source:* 2011 British Crime Survey)**

27% of respondents said they felt 'a bit unsafe' or 'very unsafe' when walking alone in their area after dark.

13% said they were 'very worried' about being burgled.

13% said they were 'very worried' about being mugged.

17% of car owners said they were 'very worried' about car theft.

In order to map fear it is necessary to draw up a scale, say 0 (no fear) to 5 (highly dangerous), that can be easily understood by people. Residents of a town might then be asked to give a score to different neighbourhoods or parts of the town on this scale. Town centres often feel safe during the day but rather threatening at night, so you may wish to specify the time of day. Make sure that, when you do your fear questionnaire, you use a stratified sample (some old people, some young, some richer, some poorer, some white, some non-white, etc.). Figure 11.35 is an example of a map of fear.

Note that it may be impossible for people to generalise their feelings about an area as big as a neighbourhood because the factors that make them fearful operate at a small scale – a particularly badly lit alleyway, for example, or an underpass that has a lot of graffiti in it.

Compare with actual local crime rates: data can be found at www.neighbourhood.statistics.gov.uk and www.police.uk

Now relate fear of crime to characteristics of the people. Do older people have a fear of crime? Or women? Or children?

Having discovered which parts of the town are more frightening to people than others, you should use a questionnaire to discover *why* this is. What are the characteristics of the area that make it frightening?

Finally, use your questionnaire to find out how fear of crime affects people's behaviour. What security measures have they taken with their homes? Is there a Neighbourhood Watch scheme? Do they stay at home in the evening because of fear of going out? Which areas would they avoid?

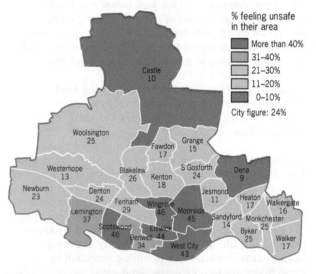

Figure 11.35  Map of Newcastle upon Tyne showing variations in reported fear of crime in 26 wards. Compare with actual local crime rates: data can be found at www.neighbourhood.statistics.gov.uk and www.police.uk

## 11.12 The geography of ill health

Since 1991 the census has identified the number of people in each area with a 'limiting long-term illness' and since 2001 asks people about their general health (very good, good, etc). The census tables break down these figures, relating health to other variables such as age, ethnicity and housing tenure.

In addition, many health authorities publish data giving the incidence of health problems by sub-area (Fig. 11.36). The 2014 Environment and Health Atlas for England and Wales gives data down to ward level.

It is therefore possible to map the distribution of ill health. Having done this, you might assess the degree of correlation with other measures. Some forms of ill health, such as heart disease or birth of low-weight babies, might correlate with low incomes. Low-income areas can be identified by using the census – they will be areas where a higher proportion of the population rent their house from the local authority.

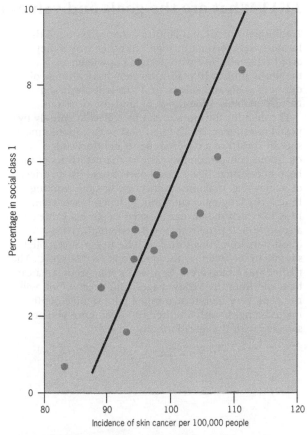

Figure 11.37  Correlation between social class and incidence of skin cancer in Croydon, south London, 2000

Some forms of ill health might correlate with prosperity – such as skin cancer, for example, which is associated with excessive sun tanning. Most forms of ill health, however, correlate with age – areas with a greater number of old people will have a greater number of residents with more serious health problems.

Such correlations can be shown by using graphs (Fig. 11.37), and the strength of the relationship should be measured using Spearman's rank correlation coefficient (page 144).

Having identified a statistical relationship, it is good practice to then look at the *reasons* for the correlation – what actually causes poorer people to have a higher incidence of heart disease, for example. In order to research this you should ask a doctor to explain the probable known causes of each type of illness you are looking at. Convert the doctor's explanation into a questionnaire that will help you to look at the lifestyle of the residents – do they smoke? Do they take exercise and if so how much? How much fresh fruit do they eat? And so on. Compare the results obtained in low-income and wealthier areas.

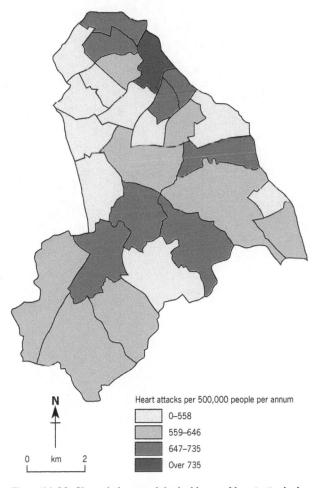

Figure 11.36  Choropleth map of the incidence of heart attacks in Croydon, south London, 2000

## 11.13 What are the goals and values of farmers?

Traditional agricultural land-use theory assumed that farmers were 'economic men', in other words they acted rationally and with perfect knowledge to maximise profits. In reality, farmers have a range of different goals and values, and this will produce a correspondingly wide range of land-use decisions.

The data for this project can be collected entirely by postal questionnaire. The first part of the questionnaire should consist of a dozen or so straightforward questions concerning the various characteristics of the farm and farmer. The second part should list a variety of values (e.g. maximising income, flexible working hours, challenging occupation, independence, etc.), next to which the farmer is asked to place a value on a scale from 0 (irrelevant) to 4 (essential). With a questionnaire of this sort, it is wise to conduct a pilot sample to see whether the questions are 'working'. The names and addresses of farmers in your study area can be taken from the Yellow Pages, and because you will not get a very high return rate a pilot sample of 50 might be right, with another 100 or so once you are satisfied with the questionnaire.

The questionnaire results will enable you to show how the goals and values of farmers vary:

a  between areas, if you conduct the survey in two different farming areas;
b  according to whether the farm is primarily arable or livestock;
c  according to farm size (acreage);
d  according to number of employees on the farm;
e  according to whether the farmer had any formal agricultural education;
f  according to the farmer's age;
g  according to farm tenure;
h  according to whether or not their parents were farmers;

or any other variables that you choose to incorporate in the questionnaire.

# *12* Cartography

When you have collected data as part of a survey or project, it will often be worth summarising in a map or diagram, because relationships between large numbers of figures can be more fully appreciated if shown in this way. This chapter discusses some of the cartographic (mapping) techniques available.

## 12.1 Graphs

Many types of graphs can be produced better by using a computer than drawing by hand. However, it is important to avoid the temptation to use the computer to graph every possible combination of variables regardless of relevance. It is also a common mistake to draw graphs with too little data to justify the exercise, such as a line graph based on only three points.

### Line graphs

Line graphs are used for showing the relationship between two variables, such as temperature and altitude (Fig. 12.1). Here are a few simple rules for drawing them.

**1** Usually one of the two variables causes the other to change (rather than vice versa). Thus if you were plotting soil depth against slope angle, it is obviously the angle of the slope that causes the soil depth to change, rather than the other way round. In this case slope angle is called the *independent variable* and goes on the horizontal or *x*-axis, while soil depth is the *dependent variable* and goes on the vertical or *y*-axis. One exception to this rule is that altitude is usually plotted on the vertical axis, regardless of whether it is the dependent or the independent variable. Thus in Figure 12.1 altitude is plotted on the *y*-axis, although it is clearly independent of temperature.

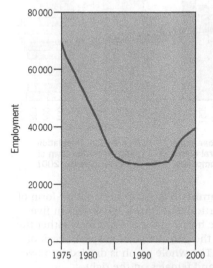

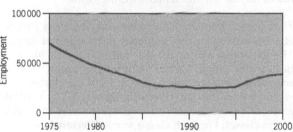

**Figure 12.2 Line graphs showing employment in London Docklands, 1975–2000**

**2** Axes should generally start at zero.
**3** Always mark on the axes what the variables are.
**4** Be aware of the fact that the scales you use on the axes will determine the visual impression given by the graph.

Figure 12.2 shows the same set of figures plotted on two graphs with axes of different scales. Note the difference in the way the two graphs look.

### Bar graphs

In a bar graph, one axis has a numerical value, but the other is simply categories. Bars are drawn proportional in height to the value they represent.

**Divided bar graphs** (Fig. 12.3) are useful when the variable being graphed (such as agricultural output) can be divided into parts (hay, turkeys, cattle). Where there is a clear pattern, the largest division of the bar should be on the bottom and the smallest on the top.

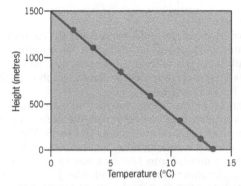

**Figure 12.1 Line graph of temperature plotted against altitude**

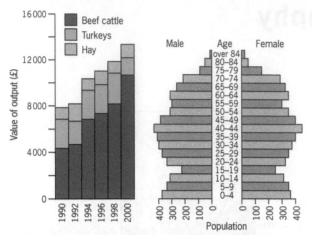

**Figure 12.3  Divided bar graph showing the agricultural output of Forest Edge Farm, Hampshire, 1990–2000**

**Figure 12.4  Population pyramid for the town of Ashdown, Sussex, 2001**

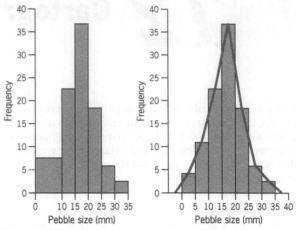

**Figure 12.6  Histogram based on the data given with Figure 12.5, combining the first two classes**

**Figure 12.7  A frequency polygon superimposed on a histogram**

**Population pyramids** (Fig. 12.4) are another form of bar graph. The vertical axis shows age groups in five-year intervals. The horizontal axis represents either the actual number or the percentage of people in each of these age groups. The whole graph is divided into two, males on the left and females on the right.

*Histograms, frequency polygons and frequency curves*
**Histograms** show the frequency distribution of data. The horizontal axis must be a continuous scale like a normal line graph, but the values marked on it represent the lower and upper limits of the classes within which the data have been grouped. The vertical axis is the frequency within which values fall into each of the classes. For each class a vertical rectangle is drawn; there must never be any gaps between the rectangles.

For example, if you have measured the lengths of a sample of 100 pebbles on a beach, you might draw a histogram similar to Figure 12.5. The horizontal axis is a series of size groups, the vertical axis is the number of pebbles that fall into each of these size groups.

The area of a histogram bar must be proportional to the frequency in the class. If we combined two classes from Figure 12.5, 0 to under 5 and 5 to under 10, giving us a total of 15 pebbles in the size class 0 to under 10, we would be wrong to draw a bar 15 units high: its area would be disproportionately large. Because area = width × height, you can adjust for a doubling of the width by halving the height (half of 15 = 7.5) (Fig. 12.6).

**Frequency polygons** You may convert a histogram into a line graph by joining the mid-points at the top of each rectangle with straight lines. The lines should be extended to a point on the horizontal axis that is half a class interval beyond the outside limits of the end classes (Fig. 12.7). This line graph is called a frequency polygon and its area is exactly equal to the area of the histogram on which it was based.

**Frequency curves** If the frequency polygon is smoothed so that there are no sharp breaks of slope, it is called a frequency curve. Such smoothing should only be done when there are a large number of classes, so that the amount of smoothing is minimal.

*Circular graphs*
These are used for plotting a variable that is continuous over a time, such as temperature data. On a normal line graph there is a false break: it starts at 1 January and ends at 31 December. On a circular graph there is no such break (Fig. 12.8).

Circular graphs are easily drawn. There are two axes – the circumference of the circle and the radius. Values (e.g. temperature) increase radially outwards. The circumference is normally time (e.g. months of the year). This is divided into 360° so a month would be 30° of the circumference (360 divided by 12). It is thus a straight-line graph stretched out and bent round.

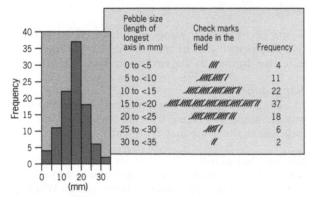

| Pebble size (length of longest axis in mm) | Check marks made in the field | Frequency |
|---|---|---|
| 0 to <5 | //// | 4 |
| 5 to <10 | ///// / | 11 |
| 10 to <15 | ///// ///// ///// // | 22 |
| 15 to <20 | ///// ///// ///// ///// ///// ///// // | 37 |
| 20 to <25 | ///// ///// /// | 18 |
| 25 to <30 | ///// / | 6 |
| 30 to <35 | // | 2 |

**Figure 12.5  Table of data about the sizes of pebbles collected on a beach, and a histogram drawn from these data**

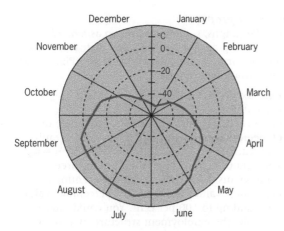

**Figure 12.8  Circular graph showing temperatures at Verkhoyansk, Siberia**

The only major problem with circular graphs is that the easily appreciated rise and fall of a normal graph is replaced by the idea of a line moving away from or towards the centre of a circle. You need considerable practice before you can appreciate its message in detail.

## Logarithmic graphs

Logarithmic graphs are of two types: those where both axes are drawn logarithmically (called log–log graphs) and those where only one axis (the vertical axis) is logarithmic (semi-log graphs). Figure 12.9(b) shows a semi-log graph. The horizontal axis is quite normal. But on the vertical scale the numbers are not spaced evenly: the interval between 20 and 30 is slightly less than between 10 and 20 and the same as between 200 and 300 higher up the scale. Numbers are regularly 'bunched', and each of these bunchings is called a *cycle*. The top and bottom of each cycle must be 10 or some decimal or multiple of 10 such as 0.1, 100, 1000 or 10 000. In each successive cycle the values are 10 times greater than those of the cycle below. Logarithmic graphs have two merits:

**1** It is possible to represent a very great range of data on one piece of graph paper. If you had to plot such values as 1, 3, 12 and 12 000, this would be impossible on a normal graph, but possible on a logarithmic graph.

**2** Equal rates of changes are shown by lines of equal slope. Compare Figures 12.9(a) and (b). Figure 12.9(a) shows a normal line graph illustrating the output from two factories over 20 years. Both factories doubled their output every 10 years (equal rates of change), but the slopes of their lines are different because factory Y has higher absolute output levels. In Figure 12.9(b) both appear as a straight line. Logarithmic graphs are therefore used for plotting rates of change.

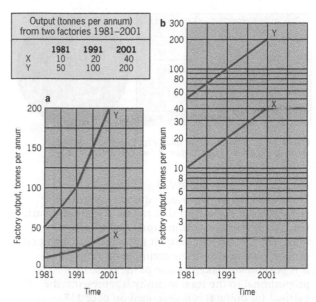

| Output (tonnes per annum) from two factories 1981–2001 | | | |
|---|---|---|---|
| | 1981 | 1991 | 2001 |
| X | 10 | 20 | 40 |
| Y | 50 | 100 | 200 |

**Figure 12.9  The same data plotted on two types of graph paper: (a) normal graph paper; (b) semi-log graph paper**

## Scatter graphs

These are used to investigate the relationship between two variables when you have data for many places. For example, Figure 12.10 shows the relationship between population size and the number of services offered in all the settlements of a region. This type of graph could be used to see if there was any relationship between birth rate and standard of living in 100 countries of the world, or between precipitation and discharge in 20 rivers.

The pattern of the scatter describes the relationship. In Figure 12.10 there is a positive correlation (as one value goes up, the other goes up) and there appear to be three main groups (similar places?). Lines are drawn on the graph that show the general trend of the dots. Three are drawn on Figure 12.10. There should be an equal number of dots above and below the line. These are called 'best-fit' lines. If the points do not form a clear trend, you should not draw a best-fit line.

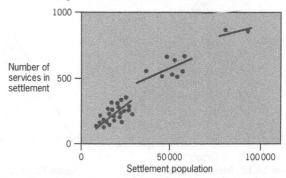

**Figure 12.10  A scatter graph: each point is a settlement**

| Population composition by ethnic origin San Fernando, Trinidad | | | |
|---|---|---|---|
| **1** Ethnic origin | **2** Total population | **3** % | **4** % of 360° |
| Black | 18784 | 47.2 | 169.9 |
| Asian | 10296 | 25.8 | 92.9 |
| White | 1306 | 3.3 | 11.9 |
| Mixed | 8283 | 20.8 | 74.9 |
| Other | 1161 | 2.9 | 10.4 |
| Total | 39830 | 100 | 360 |

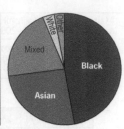

**Figure 12.11 Divided circle (pie graph), with calculations, showing the ethnic composition of San Fernando, Trinidad**

## Divided circles (pie graphs)

These are used for showing a quantity (such as the population of a country) that can be divided into parts (such as different ethnic groups). A circle is drawn to represent the total quantity. It is divided into segments proportional in size to the components (Fig. 12.11). It is possible to make the size of the circle itself proportional to the total quantity it represents; the method for doing this is described on page 137.

Unless you are using a computer to draw the pie graph for you, the method for dividing the circle is as follows:

1 Draw the circle, proportional in area to the total quantity to be represented or not, as you wish.
2 Tabulate the values that will form the segments of the circle (columns 1 and 2 in the table in Figure 12.11). Convert these into a percentage of the whole (column 3).
3 Calculate the angle which corresponds to this percentage of 360 degrees (column 4).
4 Draw a vertical line from the centre of the circle to the top of the circumference.
5 Draw in segments, measuring the angles calculated in step 3. Start from the vertical line and work in a clockwise direction. Draw the segments in order of size (largest first, etc.)
6 Different segments may be shaded differently and numbers or words written in.

## Triangular graphs

Triangular graphs, sometimes known as *ternary diagrams*, are graphs with three axes instead of two, taking the form of an equilateral triangle (Fig. 12.12). The important features are:

1 Each axis is divided into 100, representing percentages.
2 From each axis lines are drawn at an angle of 60 degrees to carry the values across the graph (Fig. 12.13).
3 The data used must be in the form of three components, each component representing a percentage value, and the three percentage values must add up to 100 per cent. You could plot, for example, the employment structure of a town:

| Employment | Percentage |
|---|---|
| Primary | 5 |
| Secondary | 33 |
| Tertiary | 62 |
| | 100 |

These figures have been plotted on Figure 12.12. Dotted lines show the way in which the values are carried across the graph until they meet at one point. The position of this point indicates the relative dominance of each of the three components. If exactly one-third of workers were engaged in each of the three employment types, the point would be in the middle of the triangle, marked X on Figure 12.12. Care must be taken when plotting and interpreting such a graph, because it can be confusing at first.

Triangular graphs can be used for plotting any data that can be conveniently divided into three portions, for example agricultural land use (arable, pastoral, other) or age structure (0–25, 26–50, over 50). Their main value arises when data for several places (or one place at several different times) are plotted on one graph. The relative position of the points then gives a quick visual impression of the relative dominance of one component (Fig. 12.14).

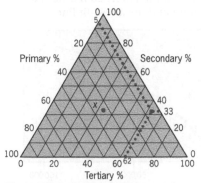
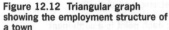

**Figure 12.12 Triangular graph showing the employment structure of a town**

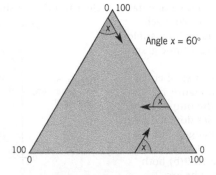

**Figure 12.13 Diagram of a triangular graph, showing the way in which the values are carried across the graph**

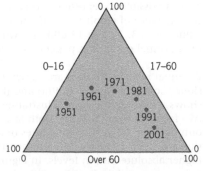

**Figure 12.14 Triangular graph of the population structure of Hatfield (a New Town), 1951–2001**

## 12.2 Isoline maps

Isolines are lines on a map that join points of equal value. We are all familiar with isolines – contours on a relief map, isotherms of temperature and isobars of pressure, for example. Isolines can only be used when the variable to be plotted changes in a fairly gradual way across space and where plenty of data are available (Fig. 12.15). If the spatial distribution is disjointed and data are not detailed enough, too much guesswork is involved in the drawing of the isolines. The method is simple:

1  Plot the data on a map as a series of points with accompanying values.
2  Decide on the interval you want between your isolines. If this is too small, there will be many isolines and the map will look cluttered. If this is too great, the map will become too generalised to be useful.
3  Draw in the isolines. You must stick to your chosen interval: all isolines should have the same interval between them. There is a good deal of personal judgement involved here. Knowledge of things like 'isobars tend to be circular around low-pressure systems but straighter between the fronts' is useful.
4  The space between different-value isolines can be shaded or coloured. The higher the value of the isolines, the darker the shading. This is the system used for portraying relief in atlases, except that by convention greens are used for low land, browns for high land, purple and white for very high mountains. If you do shade or colour, include a key. If you do not shade between isolines, mark on their numerical value.

Isolines are ideal for showing gradual change over space. They avoid the 'unreal' effect that boundary lines produce on choropleth maps (Section 12.5).

The main limitations of isolines are that they are unsuitable for 'patchy' distributions, and a large amount of data is needed to draw an accurate isoline map.

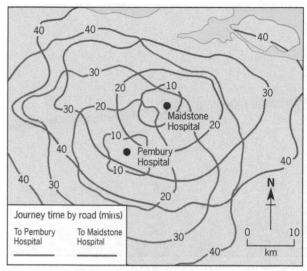

**Figure 12.15  Isochrones (lines of equal travel time) showing journey times by road to two hospitals**

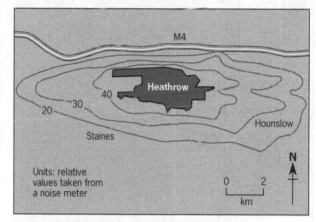

**Figure 12.16  Isoline map of aircraft noise contours around Heathrow**

## 12.3 Dot maps

In dot mapping, dots of a fixed size are given a value representing a variable, such as crop yield or numbers of people, cattle, shops, etc. Locate them on a base map roughly where that phenomenon occurs (Fig. 12.17a).

The main steps involved in the drawing of a dot map are as follows:

1  Prepare your data. If the data you are using were gathered within districts, obtain or draw a map showing the district boundaries. Remember: the larger the areal units used, the less informative the map will be.
2  Find the total number of items to be shown on the map: the number of people, cattle, etc.

3  Decide on the dot value. This should be high enough to avoid excessive overcrowding of dots in areas with a high concentration of the phenomenon being mapped, and low enough to prevent areas with low concentrations of the phenomenon having no dots at all, so giving a false impression of emptiness. It may not be possible to fulfil both these criteria, and you may have to choose a compromise value.
4  Decide on dot size. Dots should be drawn of uniform size. This size will depend on the density of dots in the area of the map which has the highest density. Ideally, dots should not be too large and should not merge.

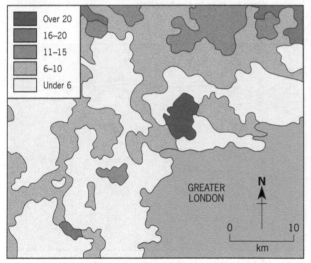

a Each dot represents 20 acres of barley

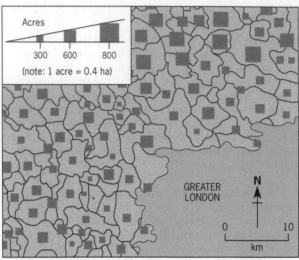

b The area of each square is proportional to the acreage of barley

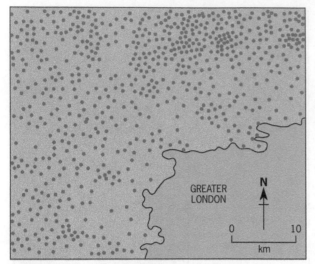

c Percentage of agricultural land under barley

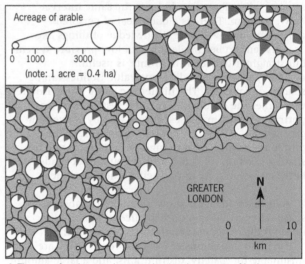

d The area of each sector is proportional to the acreage of barley

**Figure 12.17 Four maps showing the distribution of barley production in the Chilterns. The most successful is the dot map (a).**

**5** Draw dots on the map to reflect as closely as possible the distribution of the phenomenon being mapped.
**6** You can use dots of different colours. For example, if you were mapping the distribution of members of five different ethnic groups in a city, you could use five different dot colours. The same dot value should be used for each.

Dot maps are useful for showing the distribution of phenomena where values are known and a fairly accurate indication of their location is given. It is the only technique that gives this accurate indication of distribution (Fig. 12.17).

Dot maps have two limitations:

**1** Large numbers of dots are hard to count so that, while they are very good at giving an impression of distribution, they are less valuable if you need a precise idea of the values they represent.
**2** If the dots are to be plotted within areas but you know nothing about the distribution of the phenomena within these areas, other methods (such as choropleth mapping) are better. In other words, you must have a large amount of initial information before you begin to draw a dot map.

# 12.4 Proportional symbols

These symbols are drawn proportional in size to the size of the variable being represented. The symbol used can theoretically be almost anything: proportional 'men' to show military strength, proportional 'trains' to show the number of trains in an area, proportional 'factories' to represent industrial output. More common are proportional spheres and cubes, drawn three-dimensionally. But the most usual and straightforward are proportional bars, squares (Fig. 12.18) and circles (Fig. 12.19).

### Proportional bars

The method for drawing proportional bars or squares is:

**1** Examine the data and decide on your scale. The length of the bar will be proportional to the value it portrays. If the bars are too long, they get in each other's way, but if they are too short, the difference between one bar and another becomes hard to see.
**2** Draw your bars on a base map, one end of the bar located next to the place to which it refers. Bars should be of uniform width, solid-looking but not so wide that they overlap. They may be placed vertically or horizontally.
**3** Mark the scale on the map.
**4** Bars may be divided (page 131).

Proportional bars may be drawn three-dimensionally as an isometric diagram, and isometric graph paper is available to help with this. Bars are easy to draw and simple to read, but they have two limitations:

**1** Because bars are linear, it is hard to show data which have a great range.
**2** Large bars are not always visually attached to the locality they are supposed to symbolise. It is hard to get ideas of distribution from a bar map.

### Proportional circles

The area of the proportional circle is proportional to the values being mapped (Fig. 12.19). The procedure for drawing these is as follows:

**1** Calculate the square root of the values. (This is necessary because we want the area of the circle to be proportional to the values. If one value was, for example, 2 and the other was 4, drawing one circle with a radius twice that of the other would make it much more than twice the area.)
**2** Multiply each square root by a constant: this gives you the radius of each circle. The constant should be the value that will not make the largest circle too large nor the smallest too small.
**3** Draw the circles and mark the scale on the map.
**4** The circles can be divided (page 134).

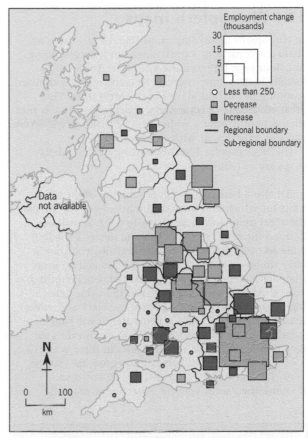

**Figure 12.18  Proportional squares: high-technology employment change, 1991–7**

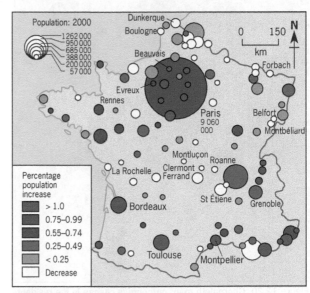

**Figure 12.19  Proportional circles: population change in the largest cities in France**

## 12.5 Choropleth maps

In choropleth or shading maps (Fig. 12.20), areas are shaded according to a prearranged key, each shading or colour type representing a range of values. It is usually best to draw choropleth maps with a computer, scanning in the boundaries of the areal units. The great advantage of computers, apart from speed, is that they generally offer a wide range of shading types. You can also experiment with different shades.

The steps involved in the drawing of choropleth maps are as follows:

**1** Obtain a base map, with boundaries of the areal units for which you have data marked on. The smaller the areal units, the more accurate the map will be.
**2** Find the range of your data and devise a shading scale accordingly. For best visual results, you should have no fewer than four shading types and no more than eight, depending on the level of detail required. Your shading should get darker as the value gets higher. It is nearly always best to use black and white shading. If you are using a computer, it will suggest a suitable range of tones for you. You can divide your data into groups of values or classes in three ways:
**i** Divide the range of values into equal-size classes: 0–4.9; 5–9.9; 10–14.9; 15–19.9; 20 and over.
**ii** Rank the values you are expecting to map and

divide the rank order into the number of groups you want. Have an equal number of values in each group. Base your shading division on the range of values represented. For example, if you have decided to use five shading types, these data would be divided into five equal or nearly equal groups:

*Percentage population change in Inner London, 1991–2001*

| City of London | −27.9 |
|---|---|
| Lambeth | −11.3 |
| Hackney | −10.0 |
| Haringey | −8.9 |
| Westminster | −7.8 |
| Kensington and Chelsea | −7.6 |
| Hammersmith and Fulham | −7.5 |
| Lewisham | −7.3 |
| Wandsworth | −6.8 |
| Southwark | −6.3 |
| Camden | −5.2 |
| Newham | −4.7 |
| Islington | −4.3 |
| Tower Hamlets | +7.5 |

**iii** Alternatively, having inspected the data, you might break them up into groups that better reflect the distribution of the data (see Figure 12.20):

*Percentage population change in Inner London, 1991–2001*

| City of London | −27.9 |
|---|---|
| Lambeth | −11.3 |
| Hackney | −10.0 |
| Haringey | −8.9 |
| Westminster | −7.8 |
| Kensington and Chelsea | −7.6 |
| Hammersmith and Fulham | −7.5 |
| Lewisham | −7.3 |
| Wandsworth | −6.8 |
| Southwark | −6.3 |
| Camden | −5.2 |
| Newham | −4.7 |
| Islington | −4.3 |
| Tower Hamlets | +7.5 |

**3** Shade in the areal units and draw a key on the map.

The choropleth shading method is easy to do and gives a good visual impression of change over space, as long as suitable shading is used. On the other hand, it suffers from two limitations:
**a** It gives a false impression of abrupt change at the boundaries of areal units. This is an unavoidable problem when a technique of this type is used.
**b** Variations *within* areal units are concealed, and for this reason small units are better than large units.

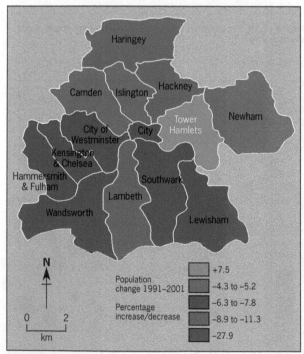

**Figure 12.20** Choropleth map: percentage change in the resident population of inner London boroughs, 1991–2001 (© Crown Copyright)

## 12.6 Flow-line maps

Flow-line maps (Fig. 12.21) are used for portraying movements or flows, such as traffic flows along roads or flows of migrants between countries. A line is drawn along the road, or from the country of origin to country of destination, proportional in width to the volume of the flow. The method for constructing a flow-line map is as follows:

**1** Draw a base map, marking relevant details such as areal units or the course of a road. If the map is to represent flows along a road network, mark in pencil the points at which vehicles were counted and the quantity of traffic counted at those points.

**2** Examine the range of your data and decide upon a scale. If the data range is not too great and the route density not too high, the scale can be a directly proportional one, for example:

1 mm thickness: 100 cars per hour
2 mm thickness: 200 cars per hour, etc.

If the flow lines are too wide, they will tend to create blocks and obscure the map. If this is the case, take the root of the value.

**3** Draw the flow lines. These may go along the actual course of the phenomenon being mapped, or direct from origin to destination, or by some other more convenient route – as long as the 'tail' of the flow line begins at the flow origin and the 'nose' of the line points towards the destination.

If the flow is a two-way movement, this can be shown by dividing a flow line and shading it, one shading type representing flow in one direction, the other representing the reverse flow (Fig. 12.22).

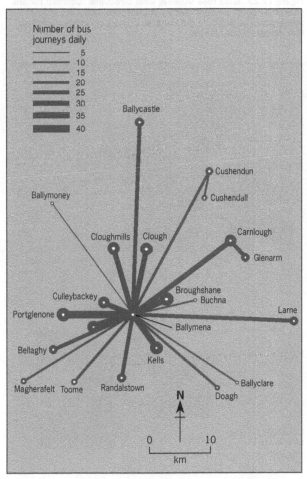

**Figure 12.21 Flow-line map showing the number of daily bus journeys from Ballymena, Northern Ireland, to surrounding settlements**

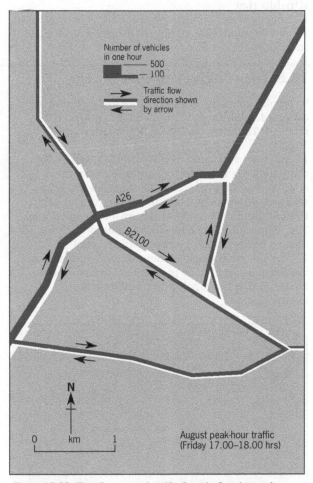

**Figure 12.22 Flow-line map of traffic flows in Crowborough, East Sussex**

## 12.7 Ray diagrams

Ray diagrams consist of straight lines (or 'rays') that show a movement or connection between two places. On any one base map, you can draw different-coloured rays to show different kinds of movement or types of connection between places. There are several types of ray diagram:

### Desire lines

A desire-line diagram shows the movement of phenomena from one place to another. Each line joins the places of origin and destination of a particular movement (see Figure 11.18 on page 117).

### Wind roses

A wind rose has rays focusing on the point from which wind direction observations have been made (Fig. 12.23). Each ray is proportional in length to the number of days in the year that wind blows from that direction.

### Kinship ties

Kinship ties (see, for example, Figure 11.3 on page 108) show two things: the number of people involved, and their location.

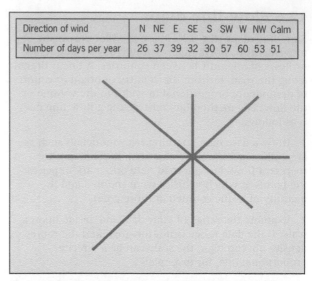

| Direction of wind | N | NE | E | SE | S | SW | W | NW | Calm |
|---|---|---|---|---|---|---|---|---|---|
| Number of days per year | 26 | 37 | 39 | 32 | 30 | 57 | 60 | 53 | 51 |

**Figure 12.23 Wind rose showing wind directions recorded for one year at a school weather station in Liverpool. Lines are drawn proportional in length to the number of days on which wind from that direction was blowing.**

## 12.8 Topological maps

In topological maps, actual distance and direction are disregarded but the relative position of places is retained. There are two types of topological map:

**1** Maps of route networks. Figure 12.24 shows how a route network might be topologically transformed: the relative position of places on the network has been retained, distances and directions ignored. The London Underground map is an example of a topological map. The reason for doing a topological transformation is to make a network map easier to read and analyse.

**2** Maps of areas (such as countries) in which the area has been distorted to be proportional to some value. Figure 12.25b is a topological transformation of Figure 12.25a.

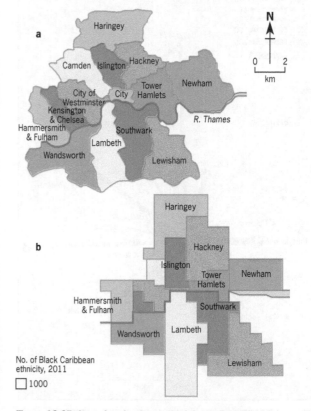

**Figure 12.25 Inner London boroughs (a), topologically transformed (b) so that boroughs are proportional in size to the number of residents of Black Caribbean ethnicity, 2011**

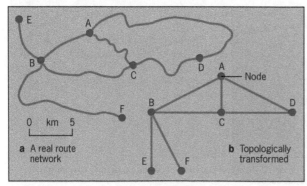

**Figure 12.24 Topological transformation**

# 13 Statistical methods

Statistical methods are used to take the analysis of data one stage beyond that which can ever be achieved with maps and diagrams. A visual inspection of data is often a vital first step, but mathematical manipulation usually gives greater precision and allows us to discover things that might otherwise go unnoticed.

Having said this, it is important to be aware that statistics are only an aid to analysis, and no more. Before you use a statistical method, it is essential to ask yourself two questions:

1 Why am I using this technique? In other words, be absolutely clear what it is you hope to prove and how this statistical method can help you do it.

2 Are my data appropriate to this particular technique? As explained below, each technique requires data to be arranged in a particular form. If they are not, the technique cannot be used. Above all, if your data are not good in the first place, the use of a complex statistical technique will not help you: 'rubbish in – rubbish out'.

In this section the working required for each technique is spelt out. However, if you have access to a calculator or computer statistics package you may well prefer to use it, as computers enable you to work faster and to explore a wider range of possibilities than working manually.

## 13.1 Mean, mode and median: descriptive statistics of central tendency

If you are faced with a large amount of data, such as the average temperature of a place every day for two years, you will probably wish to get it into a more manageable form by summarising it. This is relatively easy to do and there are three commonly used methods.

### The mean
The mean is what you know as the average, and you find it by adding together all the values under consideration and dividing the total by the number of values. The mean is shown by the symbol $\bar{x}$.

The mean is distorted if you have just one extreme value, which can be a problem. But it is the most common type of summary because it can be used for further mathematical processing.

### The mode
The mode is simply the most frequently occurring event. If we are using simple numbers, the mode is the most frequently occurring number. If we are looking at data on the nominal scale (grouped into categories), the mode is the most common category.

The mode is very quick to calculate, but it cannot be used for further mathematical processing. It is not affected by extreme values.

### The median
The median is the central value in a series of ranked values. If there is an even number of values, the median is the mid-point between the two centrally placed values.

The median is not affected by extreme values, but it cannot be used for further mathematical processing.

---

**Examples**

*Mean ($\bar{x}$)*

Data: 3, 4, 4, 4, 6, 6, 9

Mean: $\dfrac{3 + 4 + 4 + 4 + 6 + 6 + 9}{7} = \dfrac{36}{7} = 5.1$

$\bar{x} = 5.1$

*Mode*

Data: 3, 4, 4, 4, 6, 9
Mode (most frequently occurring number) = 4

*Median*

Data: 3, 4, 4, 4, 6, 9
Median (central value) = 4

Data: 3, 4, 4, 6, 6, 9
Median = 5

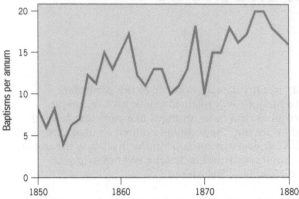

Figure 13.1(a) **Graph of baptisms recorded at the Church of St Bede, Jarrow, 1850–80**

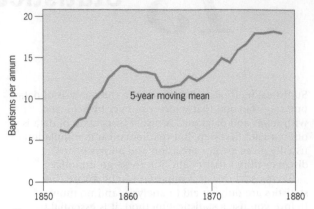

Figure 13.1(b) **Graph of baptisms recorded at the Church of St Bede, Jarrow, 1850–80, five-year moving means**

*Moving means*

Figure 13.1(a) shows the number of baptisms in a parish over a thirty-year period. Such data often contain a large number of short-term random fluctuations that make it hard to identify longer-term trends. These short-term fluctuations can be adjusted on the graph by calculating and plotting moving means. The degree of ironing-out required and the number of years for which you have data will determine the time interval chosen for the purposes of this method. Figure 13.1(b) was drawn after

calculating five-year moving means, but if the graph had been for the period 1850–1950, it might have been better to use eleven-year moving means (always use an odd number).

To find a five-year moving mean, calculate the average figure for the first five years taken together (1850, 1851, 1852, 1853, 1854) and plot the result under the third year (1852). You then find the average for years two to six (1851, 1852, 1853, 1854, 1855) and plot the figure for the middle of these (1853), and so on for all years in groups of five.

## 13.2 Spread around the median and mean: descriptive measures of difference

The median, mean and mode all give us a summary value for a set of data. On their own, however, they give us no idea of the spread of data around the summary value, which can be misleading. For example, if we were looking at rainfall figures for a semi-arid area, the data might read as follows:

| Year | Rainfall (mm) |
|------|---------------|
| 1990 | 0 |
| 1991 | 0 |
| 1992 | 3 |
| 1993 | 0 |
| 1994 | 97 |

The mean for this data (20 mm) gives an untrue picture of what really happened. There is a very great 'deviation about the mean'. The median (0 mm) is similarly misleading – the deviation is even greater.

Deviation can be measured statistically as follows.

## Spread around the median: the interquartile range

The interquartile range is a measure of the spread of values around their *median*. The greater the spread, the higher the interquartile range.

### Method

Stage 1    Place the variables in rank order, smallest first, largest last.

Stage 2    Find the upper quartile. This is found by taking the 25 per cent highest values and finding the mid-point between the lowest of these and the next-lowest number.

Stage 3    Find the lower quartile. This is obtained by taking the 25 per cent lowest values and finding the mid-point between the highest of these and the next-highest value.

Stage 4    Find the difference between the upper and lower quartiles. This is the interquartile range, a crude index of the spread of values around the median. The higher the interquartile range, the greater the spread.

## Spread about the mean: the standard deviation ($\sigma$)

If we want to obtain some measure of the spread of our data around its mean, we calculate its standard deviation. Two sets of figures can have the same mean but very different standard deviations.

### Method

Stage 1    Tabulate the values ($x$) and their squares ($x^2$). Add these values ($\Sigma x$ and $\Sigma x^2$).

Stage 2    Find the mean of all the values of $x$ ($\bar{x}$) and square it ($\bar{x}^2$).

Stage 3    Calculate the formula:

$$\sigma = \sqrt{\left(\frac{\Sigma x^2}{n} - \bar{x}^2\right)}$$

where   $\sigma$ = standard deviation
$\sqrt{}$ = square root of
$\Sigma$ = the sum of
$n$ = the number of values
$\bar{x}$ = the mean of the values

The higher the standard deviation, the greater the spread of data around the mean. The standard deviation is the best of the measures of this spread, because it takes into account all the values under consideration.

### Example

*Monthly average temperatures (°C), Middleton, Norfolk*

| | | | |
|---|---|---|---|
| January | 4 | July | 17 |
| February | 5 | August | 17 |
| March | 7 | September | 15 |
| April | 9 | October | 11 |
| May | 12 | November | 7 |
| June | 15 | December | 5 |

Ranked: 4   5   5   7   7   9   11   12   15   15   17   17

lower quartile 6

upper quartile 15

Interquartile range: $(15 - 6) = 9$

### Example

*Number of vehicles passing a traffic count point on ten days between 9.00 and 10.00 a.m.*

| | | |
|---|---|---|
| Day 1 | 50 | Day 6   70 |
| Day 2 | 75 | Day 7   63 |
| Day 3 | 80 | Day 8   42 |
| Day 4 | 92 | Day 9   75 |
| Day 5 | 60 | Day 10   82 |

If we were planning a traffic management scheme to cope with the traffic on this road, we would not only be interested in the average (or mean) amount of traffic that passes along that road, but also in the amount of traffic on particularly busy and idle days of the week. The greater the standard deviation, the greater the difference in volume between these high and low periods. If the standard deviation were very high, we might wish to install traffic lights that only operated at the busiest times of the day or week.

| $x$ | $x^2$ |
|---|---|
| 50 | 2 500 |
| 75 | 5 625 |
| 80 | 6 400 |
| 92 | 8 464 |
| 60 | 3 600 |
| 70 | 4 900 |
| 63 | 3 969 |
| 42 | 1 764 |
| 75 | 5 625 |
| 82 | 6 724 |
| $\Sigma x = 689$ | $\Sigma x^2 = 49\ 571$ |

$$\bar{x} = \frac{689}{10} = 68.9 \qquad \bar{x}^2 = (68.9)^2 = 4747.2$$

$$n = 10$$

$$\sigma = \sqrt{\left(\frac{\Sigma x^2}{n} - \bar{x}^2\right)} = \sqrt{\left(\frac{49\ 571}{10} - 4747.2\right)} = 14.5$$

## 13.3 Correlation: descriptive measures of association

Two things *correlate* when they vary together, such as temperature decreasing with altitude or land values falling with distance from the city centre. If, as one variable increases in value so does the other, this is a *positive* correlation. If one goes up as the other goes down, this is a *negative* correlation.

Some things correlate fairly exactly: temperature falling with altitude, for example. Other things correlate but not very well, such as people's age and height. These correlations can be seen if the two variables are plotted on a graph (Fig. 13.2). The level of correlation can also be expressed as a numerical index, and this is what the following technique does: it expresses a relationship as a number, known as a *correlation coefficient*. This is useful for three reasons.

**1** It is more precise than a graph. While two graphs showing correlations may look similar, the correlation coefficients for the sets of data may well be slightly different.
**2** If we wanted to compare several pairs of data, such as the relationship between temperature and altitude on twenty slopes, it would be far easier to compare twenty *numbers* than twenty *graphs*.
**3** It is possible to test the correlation to see if it is really significant or whether it could have occurred by chance alone.

### A warning
The fact that two things correlate proves nothing. We can never conclude from statistical evidence alone that, because two things correlate, one must be affecting the other. If we found that crop yields decreased with height up a mountain, we would not know without further research whether this was owing to the fall in temperature, or steeper slopes, or changes in precipitation. All projects involving correlations must be supplemented by research that seeks to uncover the processes behind the correlation.

### *Spearman's rank correlation coefficient ($r_s$)*
This technique is among the most reliable methods of calculating a correlation coefficient. This is a number which will summarise the strength and direction of any correlation between two variables.

**Method**

*Stage 1*  Tabulate the data as shown in the example. Rank the two data sets independently, giving the highest value a rank 1, and so on.

*Stage 2*  Find the difference between the ranks of each of the paired variables ($d$). Square these differences ($d^2$) and sum them ($\Sigma d^2$).

*Stage 3*  Calculate the coefficient ($r_s$) from the formula:

$$r_s = 1 - \left( \frac{6\Sigma d^2}{n^3 - n} \right)$$

where  $d$ = the difference in rank of the values of each matched pair
  $n$ = the number of pairs
  $\Sigma$ = the sum of

The result can be interpreted from the scale:

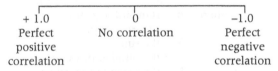

| + 1.0 | 0 | −1.0 |
|---|---|---|
| Perfect positive correlation | No correlation | Perfect negative correlation |

You should now determine whether the correlation you have calculated is really significant, or whether it could have occurred by chance.

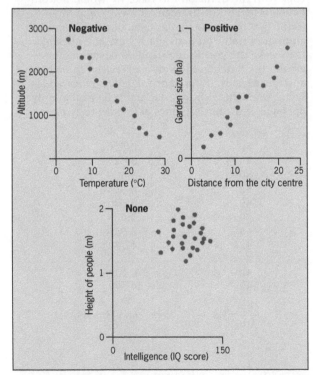

**Figure 13.2  Types of correlation**

*Stage 4* Decide on the rejection level ($\alpha$). This is simply how certain you wish to be that the correlation you have calculated could not just have occurred by chance. Thus, if you wish to be 95 per cent certain, your rejection level is calculated as follows:

$$\alpha = \frac{100 - 95}{100}$$
$$= 0.05$$

*Stage 5* Calculate the formula for *t*:

$$t = r_s \sqrt{\left(\frac{n - 2}{1 - r_s^2}\right)}$$

where  $r_s$ = Spearman's rank correlation coefficient
  $n$ = number of pairs

*Stage 6* Calculate the degrees of freedom (df):
  (df) = $n - 2$

where  $n$ = the number of pairs

*Stage 7* Look up the critical value in the *t*-tables (page 161), using the degrees of freedom (df, stage 6) and rejection level ($\alpha$, stage 4). If the critical value is less than your *t*-value (stage 5), then the correlation is significant at the level chosen (95 per cent). If the critical value is more than your *t*-value, then you cannot be certain that the correlation could not have occurred by chance. This may mean one of two things:

**a** The relationship is not a good one and it is thus not really worth pursuing it any further.
**b** The size of sample you are using is too small to permit you to prove a correlation. If you increase the sample size, a statistically significant correlation may then emerge.

It is not possible to tell which of these conclusions is the correct one from the technique alone. It requires intelligent geographical thinking on your part to decide this.

**Example**

Population size and number of services in each of 12 settlements:

| Population | No. of services |
|---|---|
| 350 | 3 |
| 5 632 | 41 |
| 6 793 | 43 |
| 10 714 | 87 |
| 220 | 4 |
| 15 739 | 114 |
| 8 763 | 72 |
| 7 982 | 81 |
| 6 781 | 73 |
| 4 981 | 35 |
| 1 016 | 11 |
| 2 362 | 19 |

*Stages 1–2*

| Settlement population | Rank | No. of services | Rank | Difference between ranks (d) | $d^2$ |
|---|---|---|---|---|---|
| 220 | 12 | 4 | 11 | 1 | 1 |
| 350 | 11 | 3 | 12 | 1 | 1 |
| 1 016 | 10 | 11 | 10 | 0 | 0 |
| 2 362 | 9 | 19 | 9 | 0 | 0 |
| 4 981 | 8 | 35 | 8 | 0 | 0 |
| 5 632 | 7 | 41 | 7 | 0 | 0 |
| 6 781 | 6 | 73 | 4 | 2 | 4 |
| 6 793 | 5 | 43 | 6 | 1 | 1 |
| 7 982 | 4 | 81 | 3 | 1 | 1 |
| 8 763 | 3 | 72 | 5 | 2 | 4 |
| 10 714 | 2 | 87 | 2 | 0 | 0 |
| 15 739 | 1 | 114 | 1 | 0 | 0 |

$$\sum d^2 = 12$$

*Stage 3*  $r_s = 1 - \left(\frac{6\sum d^2}{n^3 - n}\right) = 1 - \left(\frac{6 \times 12}{12^3 - 12}\right)$

$= + 0.96$ (a strong positive correlation)

*Stage 4*  Rejection level ($\alpha$) = 95%
  = 0.05

*Stage 5*  $t = r_s \sqrt{\left(\frac{n - 2}{1 - r_s^2}\right)} = 0.96 \sqrt{\left(\frac{12 - 2}{1 - 0.96^2}\right)} = 10.73$

*Stage 6*  df = ($n - 2$) = (12 – 2) = 10

*Stage 7*  df = 10
  rejection level = 0.05
  therefore critical value of $t$ = 2.23

The critical value is less than our *t*-value (10.73). We can therefore conclude that there is a significant correlation between settlement size and the number of services offered in each.

# 13.4 Confidence limits

In a research project you usually collect sample data: a sample of 100 people, 100 houses or 100 stream measurements. It is helpful to estimate how close the results you get from measuring your samples are to the result you would get if you measured the total population (all people, all houses, all points on the stream). These are *confidence limits*.

## Calculating confidence limits

### 30 or more samples
If we wished to estimate the length of pebbles on a beach, the method would be as follows:

*Stage 1*   Take a random sample: 100 pebbles ($n$)
*Stage 2*   Mean length of this sample of pebbles: 50 mm ($x$)
*Stage 3*   Standard deviation of this data (above): 10 ($s$)
*Stage 4*   Calculate the standard error of the sample mean (SE$\bar{x}$):

$$\text{SE}\bar{x} = \frac{s}{\sqrt{n}} = \frac{10}{\sqrt{100}} = 1$$

From this simple calculation we can now make the following probability statements:

a   There is a 68 per cent probability that the mean length of all pebbles lies within one standard error of our sample mean, i.e. 49–51 mm.
b   There is a 95 per cent probability that the real mean is within two standard errors of our sample mean, i.e. 48–51 mm.
c   There is a 99.7 per cent probability that the real mean is within three standard errors of our sample mean, i.e. 47–53 mm.

The percentage probability figure is known as the *confidence level*. The range of values within which the real mean might lie are *confidence limits*.
   For values other than 68, 95.7 and 99.77 per cent:

1   Halve your required confidence level and divide by 100, e.g. 90 per cent = 0.45.
2   Look this up in Appendix A2, column A. Find the corresponding $z$ value (1.6).
3   Confidence limits are then:
    = $\bar{x} \pm (z \times$ the standard error (SE))
    = 50 ± (1.6 × 1) = 48.4 – 51.6

### Under 30 samples
*Stage 1*   Find the best estimate of standard deviation ($\hat{\sigma}$).

$$\hat{\sigma} = s \sqrt{\left(\frac{n}{n-1}\right)}$$

where $n$ = sample size
      $s$ = sample standard deviation (page 143)

e.g. sample size       = 20 pebbles
     mean pebble size  = 41 mm
     standard deviation
     of lengths        =  9 mm

$$\hat{\sigma} = 9 \sqrt{\left(\frac{20}{20-1}\right)} = 9.2$$

*Stage 2*   Find the standard error of sampling distribution (SE$\bar{x}$)

$$\text{SE}\bar{x} = \frac{\hat{\sigma}}{\sqrt{n}}$$

$$= \frac{9.2}{\sqrt{20}}$$

$$= 2.0$$

*Stage 3*   Look up the $t$-value in Appendix A3.

In this table, *degrees of freedom* are the sample size ($n$) minus one = 20 – 1 = 19.
   The rejection level is something you must decide upon – it depends how certain you wish to be. $p = 0.05$ means you will be 95 per cent certain of your final result (Stage 4 below). $p = 0.01$ means you will be 99 per cent certain. (You always multiply the $p$ value by 100 and then take it away from 100.) If we choose 95 per cent, looking up my figure 19 in column 0.05, my result is 2.09.

*Stage 4*   Multiply SE$\bar{x}$ by the $t$-value.

$$2 \times 2.09 = 4.18$$

We can now say that my mean pebble size is 41 ± 4.18 mm at the 95 per cent level of probability (37–45 mm). In other words, if I picked up every pebble on the beach and measured its length, there is a 95 per cent likelihood that the average length we would find would be something between 37 mm and 45 mm.

# 13.5 Tests of significance

When carrying out a project you will often collect two or more sets of sample data with the aim of comparing them and demonstrating contrasts. Examples might include:

1 land values at the centre and outskirts of a town;
2 temperatures in and out of a wood;
3 pebble sizes at either end of a beach;
4 crop yields on two different rock types.

Tests of significance are used to tell us whether the differences between two or more sets of sample data are truly significant or whether these differences could have occurred by chance. For example, if we measured the temperatures twenty times on a north-facing slope and twenty times on a south-facing slope the result might be: north-facing 13.4 °C; south-facing 13.7 °C.

Can we now say with confidence that the actual (rather than sampled) temperatures are higher on the south-facing slope? Or could it be that differences between the figures are due to chance and that another sample would give a different result? Tests of significance tell us the probability that differences between sample data are due to chance.

If we find that there is a significant probability that the relationship could have occurred by chance, this can mean one of two things:

1 The relationship is not significant and there is little point in looking further for explanations of it.
2 Our sample is too small. If we took a larger sample, we might well find that the result of the test of significance changes: the relationship becomes more certain.

It is not possible to tell which of these conclusions is the correct one from the result of the test itself. This is a good example of the way that statistics are only a tool and can never replace good geographical thinking.

## The Mann–Whitney U test

This test should be used when you are comparing two places, such as the microclimate of a wood compared with a field. There are three versions. Which you use depends on the number of observations you have made.

### Fewer than nine readings at each place

*Stage 1* Call one sample A and the other B.
*Stage 2* Call the number of values in the smaller sample $n_1$, and the number of values in the larger sample $n_2$.
*Stage 3* Place all the values together in rank order (i.e. from lowest to highest).
*Stage 4* Inspect each B in turn and count the number of As that precede it. Add up the total to get a U value.
*Stage 5* Repeat stage 4, but this time inspect each A in turn and count the number of Bs that precede it. Add up the total to get a second U value.
*Stage 6* *Taking the smaller of the two U values*, look up the probability value associated with it in the appropriate table of Appendix A4. Multiply this by 100 to get the percentage probability that the difference between your two sample sets could have occurred by chance.

---

### Example

*Stage 1*
A Temps out of wood (°C): 3.5; 3.7; 4.4; 4.0; 4.6; 4.5
B Temps in wood (°C):  3.2; 3.3; 4.2; 3.4; 3.6; 4.3

*Stage 2* $n_1 = 6$
  $n_2 = 6$

*Stage 3*

| B | B | B | A | B | A | A | A | B | B | A | A |
|---|---|---|---|---|---|---|---|---|---|---|---|
| 3.2 | 3.3 | 3.4 | 3.5 | 3.6 | 3.7 | 4.0 | 4.2 | 4.3 | 4.4 | 4.5 | 4.6 |

*Stage 4* $U = 0 + 0 + 0 + 1 + 3 + 3 = 7$

*Stage 5* $U = 3 + 4 + 4 + 6 + 6 + 6 = 29$

*Stage 6* $U = 7$. The correct table is D. The critical value $= 0.047 \times 100 = 4.7$

---

The chance that the temperatures measured out of the wood were warmer just by chance is only 4.7 per cent.

**Fewer than 20 readings ($n_1$) in one sample and 9–20 readings ($n_2$) in the other**

*Stage 1*  Tabulate the readings in two separate columns. Rank the readings, treating them as one group. The smallest value is rank 1.

*Stage 2*  Add the ranks for $n$, (to get $\Sigma R_1$). Add the ranks for $n_2$ (to get $\Sigma R_2$).

*Stage 3*  Calculate the formula:
$$U_1 = n_1 n_2 + \tfrac{1}{2} n_1 (n_1 + 1) - \Sigma R_1$$

where $n_1$ = number of sample readings in one area
$n_2$ = number of sample readings in the other area, with $n_1$ representing the smaller of the two numbers, if they are different.
$\Sigma R_1$ = sum of the ranks of readings $n_1$

*Stage 4*  Calculate the similar formula:
$$U_2 = n_1 n_2 + \tfrac{1}{2} n_2 (n_2 + 1) - \Sigma R_2$$

where $\Sigma R_2$ = sum of the ranks of readings $n_2$.

*Stage 5*  You now have two $U$ values. Select the smallest of the two.

Using your values of $n_1$ and $n_2$, look up the critical value of $U$ in Table J, Appendix A4.

If the critical value of $U$ is more than your calculated value of $U$ then there is a 5 per cent or less probability that the difference between the two sample sets could have occurred by chance.

If the critical value of $U$ is less than your calculated value of $U$, then there is a more than 5 per cent probability that the difference between the two sample sets could have occurred by chance alone.

**Example**

*Aim:* to see if office rents are higher in west London than London Docklands.

*Stages 1–2*

| Rents per square foot | | | |
|---|---|---|---|
| Ten new office blocks in west London ($n_1 = 10$) | | Thirteen office blocks in Docklands ($n_1 = 13$) | |
| £ | Rank ($R_1$) | £ | Rank ($R_2$) |
| 32.6 | 13 | 30.4 | 6 |
| 31.2 | 11 | 26.6 | 2 |
| 36.6 | 23 | 31.0 | 9 |
| 35.6 | 20 | 31.1 | 10 |
| 36.1 | 21 | 32.0 | 12 |
| 28.4 | 4 | 33.2 | 15 |
| 34.5 | 18 | 28.3 | 3 |
| 34.2 | 17 | 32.9 | 14 |
| 34.9 | 19 | 30.9 | 8 |
| 33.9 | 16 | 30.0 | 5 |
| | | 36.4 | 22 |
| | | 30.6 | 7 |
| | | 26.2 | 1 |
| $\Sigma R_1 = 162$ | | $\Sigma R_2 = 114$ | |

*Stage 3*  $\begin{aligned} U_1 &= n_1 n_2 + \tfrac{1}{2} n_1 (n_1 + 1) - \Sigma R_1 \\ &= 10 \times 13 + \tfrac{1}{2} \times 10 (10 + 1) - 162 \\ &= 23 \end{aligned}$

*Stage 4*  $\begin{aligned} U_2 &= n_1 n_2 + \tfrac{1}{2} n_2 (n_2 + 1) - \Sigma R_2 \\ &= 10 \times 13 + \tfrac{1}{2} \times 13 (13 + 1) - 114 \\ &= 107 \end{aligned}$

*Stage 5*  $\begin{aligned} n_1 &= 10 \\ n_2 &= 13 \end{aligned}$

Critical value = 37

37 is more than 23, therefore there is a 5 per cent or smaller probability that property values are higher in west London than in the Docklands.

**Samples of more than 20**

*Stages 1–3*  as above. This will give you the value of $U$, $n_1$ and $n_2$.

*Stage 4*  Calculate $z$
$$z = \frac{U - \tfrac{1}{2} n_1 n_2}{\sqrt{\left( \frac{(n_1)\,(n_2)\,(n_1 + n_2 + 1)}{12} \right)}}$$

*Stage 5*  Look up your critical value in Appendix A2, column C, using your $z$ value. Multiply this by 100 to find the percentage probability that the difference between the two sample sets could be due to chance.

## The chi-squared ($\chi^2$) test

The chi-squared test can only be used on data that has the following characteristics:

1  The data must be in the form of frequencies counted in each of a number of categories. Data on the interval scale (data that have precise numerical meanings, such as height above sea level, the population size of a town or temperature) can be grouped into categories to enable you to use this test.
2  The total number of observations should be greater than 20.
3  The expected frequency in any one fraction (described in stage 3 below) should not be less than 5.
4  The observations should not be such that one influences another.

### Chi-squared for two variables

*Stage 1*  Tabulate the data as shown in the example. The data being tested for significance are called the *observed frequency*, and the column is thus headed O.

*Stage 2*  Calculate the number of counts you would *expect* to find in each category if the categories had no impact on these. This is the *expected frequency (E)*.

*Stage 3*  Calculate the formula:

$$\chi^2 = \Sigma \frac{(O - E)^2}{E}$$

where $\chi^2$ = chi-squared figure
$\Sigma$ = the sum of
$O$ = observed frequency
$E$ = expected frequency

*Stage 4*  Calculate degrees of freedom. This is simply one less than the total number of categories:
df = $n - 1$

where df = degrees of freedom
$n$ = number of categories in the test

*Stage 5*  Turn to Appendix A5 and, using the calculated value of $\chi^2$ and the degrees of freedom, read off the probability that the data frequencies you are testing could have occurred by chance.

### Example

You have visited four equal-sized areas, each on a different rock type. You counted the number of streams on each area. The results were as follows:

| Rock type | No. of streams |
|---|---|
| Chalk | 7 |
| Granite | 58 |
| Limestone | 15 |
| Sandstone | 20 |

You now wish to know if these results are a true reflection of the nature of each rock type or whether they could simply be the result of chance.

The $\chi^2$ test can be used because:

a  The data is in the form of counts.
b  The total number of streams observed exceeds 20.
c  The expected frequency in any one fraction exceeds 4. This is the number you would *expect* if rock type had no influence on stream densities. In this case it is the total number of streams (100) divided by the number of rock types (4).
d  The observations are independent (the number of streams on one rock type does not influence the number of streams on another).

The hypothesis we are trying to prove is that there is a significant difference between the sample data sets, i.e. rock type does influence stream density.

*Stage 1*  Tabulation: see stage 3.

*Stage 2*  If rock type had no influence on stream density, we would expect an equal number of streams on each rock. There are 100 streams. The areas examined on each rock type are the same. You would therefore expect 25 streams on each rock type, $E = 25$. If the areas surveyed on each rock type were not equal, the expected frequencies would have to be divided among the rock types proportional to their areas.

*Stage 3*

| Rock type | Observed frequency (O) | Expected frequency (E) | $\dfrac{(O-E)^2}{E}$ |
|---|---|---|---|
| Chalk | 7 | 25 | $(7-25)^2 \div 25 = 13.0$ |
| Granite | 58 | 25 | $(58-25)^2 \div 25 = 43.6$ |
| Limestone | 15 | 25 | $(15-25)^2 \div 25 = 4.0$ |
| Sandstone | 20 | 25 | $(20-25)^2 \div 25 = \underline{1.0}$ |
| | | | $\Sigma\ 61.6$ |

$$\chi^2 = \Sigma\frac{(O-E)^2}{E} = 61.6$$

*Stage 4*  $\begin{aligned} df &= n - 1 \\ &= 4 - 1 \\ &= 3 \end{aligned}$

*Stage 5* From the graph in Appendix A5 read off the degrees of freedom (3) on the horizontal axis against the $\chi^2$ value (61.6) on the vertical axis. The resulting point is above the line marked 0.1 chance in 100. This means that the probability that the data given above could be due to chance alone is less than 1 in 1000.

# *14* Spatial analysis

If we are interested in the distribution of phenomena such as settlements, shops or landforms, we might use spatial analysis. Spatial analysis enables us to give quantitative (numerical) measures to these things and thus adds a level of precision that simple verbal description cannot give. It may also help us to discover something about the distribution that was not immediately obvious.

## 14.1 Analysis of point distributions

The following methods of analysis are applicable to any geographical feature that forms a point distribution (rather than a line or area), such as towns, factories, shops, drumlins or erratics.

### *The median centre*
This method locates the central point of a distribution. This is useful when comparing the distribution of two or more things, such as bookshops compared with greengrocers, because it enables us to see the patterns more clearly.

### Method

*Stage 1* Plot the points on a map.
*Stage 2* Draw two lines, running north–south and east–west respectively, such that they bisect the point distribution, leaving an equal number of points on either side of each line. If the total number of points is an odd number, the lines must pass through a point. If the total is an even number, the lines must be halfway between two points. Where the lines intersect is the central point, known as the *median centre*.

Figure 14.1 shows the median points calculated for the distribution of stockbrokers, banks and commodity brokers in the City of London. It shows, in a simple summary form, how each of these three has a different centre of gravity within the City. Seeing how these median centres have changed over the past twenty years would be an interesting study in its own right.

### *Distribution of points around the median centre: the index of dispersion*
It is often useful to see the degree to which points, such as banks or stockbrokers, are clustered in a tight group or fairly widely dispersed. The index of dispersion will tell us this.

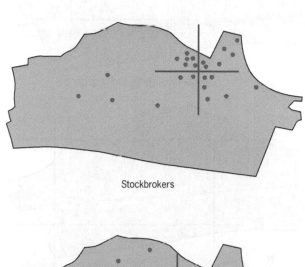
Stockbrokers

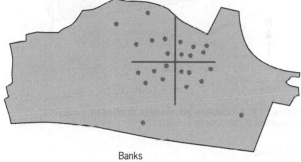
Banks

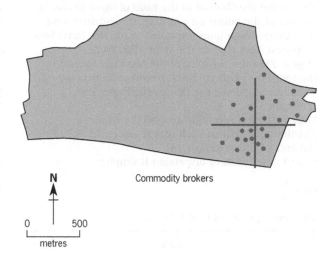
Commodity brokers

N
0    500
metres

**Figure 14.1 Median points of the distribution of stockbrokers, banks and commodity brokers in the City of London**

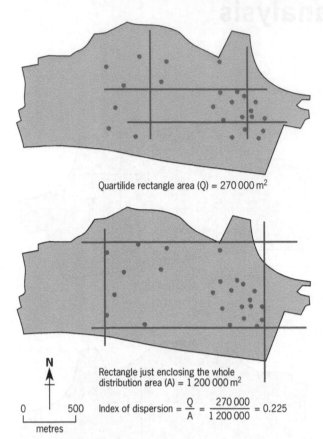

Quartilide rectangle area (Q) = 270 000 m²

Rectangle just enclosing the whole
distribution area (A) = 1 200 000 m²

Index of dispersion = $\dfrac{Q}{A} = \dfrac{270\,000}{1\,200\,000} = 0.225$

**Figure 14.2  Calculating the index of dispersion: insurance companies in the City of London**

## Method

*Stage 1* Plot the points on a map.
*Stage 2* Draw two north–south lines, each of which divides the distribution in the ratio of three to one (a quarter of the points on one side, three-quarters on the other). These lines are called *quartilides*. Draw two east–west lines that do the same (Fig. 14.2).
*Stage 3* Measure the area of the rectangle formed by the four lines. This is called the *quartilide rectangle* (Q). The more dispersed the points, the bigger the rectangle.
*Stage 4* Draw a rectangle around the whole distribution of points such that it just encloses it. Find the area of this rectangle (A).
*Stage 5* The index of dispersion is simply:

$$\text{ID} = \frac{Q}{A}$$

The range of values is as follows:

| | | |
|---|---|---|
| 0 | 0.25 | 1 |
| highly clustered | uniform | completely dispersed |

## Advantages

As with all such indices, there is little point to them if you have just one distribution to analyse: converting the pattern to an index will tell you no more than you can see from the map of the distribution. The index becomes useful in two situations.

**1** If you have two or more point distributions that are quite similar to each other. Comparing maps of their distributions may not allow you to decide which is the more clustered. Calculating the index will bring out such fine distinctions.
**2** If you had to compare a large number of point distributions, such as the pattern of settlements in fifty different areas of the UK, it would be very hard to do this by using maps alone. However, if we calculated the index of dispersion for each map, we could rank every one of the fifty areas according to how concentrated its settlement pattern was. In this way we can compare every map with every other map, as well as doing further mathematical manipulation such as calculating the average index.

## Disadvantages

The disadvantage of the index of dispersion is that it is possible for the points to be dispersed relative to the median centre, scoring a high index, but to be grouped in small clusters in sites around the edge of the distribution. The fact that they are grouped relative to each other would not be revealed by the index.

## *Distribution of points relative to each other: nearest neighbour index*

The nearest neighbour index does the same job as the index of dispersion: it summarises, in one number, how clustered, uniform or random the distribution of a series of points is. The only difference is that the degree of clustering is measured by looking at the relationship of the points to each other rather than to a central point. This can sometimes make it a more meaningful index than the index of dispersion.

## Method

*Stage 1* Plot the points on a map and number them.
*Stage 2* Draw up a table as shown and find the distance from each point to its nearest neighbouring point. It is quite possible that a point will be the nearest neighbour of several other points, or that two points will be each other's nearest neighbour.
*Stage 3* Add these nearest-neighbour distances and divide the result by the number of points (*n*). This gives the mean observed distance of all points to their nearest neighbours ($\overline{D}$).

*Stage 4* Calculate the nearest neighbour index from the following formula:

$$NNI = 2\overline{D}\sqrt{\frac{n}{A}}$$

where NNI = nearest neighbour index
$\overline{D}$ = mean observed nearest neighbour distance
$\sqrt{}$ = square root of
$n$ = total number of points
$A$ = area of the map on which the points lie

The units used for this must be the same as those used to calculate distances in stage 2 above. The result can be interpreted from the following scale:

```
0                    1.0                   2.15
completely        completely          completely
clustered           random              uniform
```

If the index value were, say, 0.7, then we would express this in words as being more nearly random than clustered. If it were 1.9, it would be more nearly uniform than random.

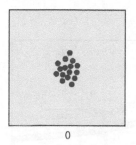

0

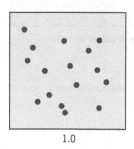

1.0

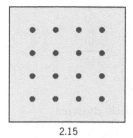

2.15

**Figure 14.3  Scale of nearest neighbour index values**

## Example

| Point | Nearest neighbour | Distance apart (km) |
|-------|-------------------|---------------------|
| 1 | 2 | 2.5 |
| 2 | 1 | 2.5 |
| 3 | 2 | 5.0 |
| 4 | 5 | 4.0 |
| 5 | 6 | 3.0 |
| 6 | 5 | 3.0 |
| 7 | 8 | 1.0 |
| 8 | 7 | 1.0 |
| 9 | 10 | 1.0 |
| 10 | 9 | 1.0 |
| | | 24.0 |

**Figure 14.4  Calculating the nearest neighbour index**

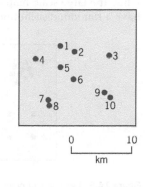

$\Sigma D = 24$
$n = 10$

$\overline{D} = \dfrac{24}{10}$

$= 2.4$ km
$A = 256$ km$^2$

$NNI = 2\overline{D}\sqrt{\dfrac{n}{A}} = 2 \times 2.4 \sqrt{\dfrac{10}{256}}$

$= 2 \times 2.4 \times 0.2$

$= 0.96$ (an almost random distribution)

## Possible problems

There are a number of problems with the nearest neighbour index, and you should be aware of these.

**1** It cannot distinguish between a single and a multi-clustered pattern. Both the distributions in Figure 14.5, although different, have an NNI of approximately 0.

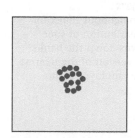
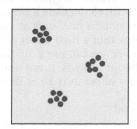

**Figure 14.5  Single and multi-clustered point patterns**

**2** An index of 1.0 does not always mean that the distribution is totally random. Two sub-patterns on the map, when combined in one index, may give a false impression of randomness. Thus Figure 14.6 produces an NNI of 1.0, although it is clearly not random.

**3** A distribution with an NNI of 1.0 should not be interpreted as being caused by random or chance factors. A pattern of settlements might have an NNI of 1.0, but every settlement is located at the site of a spring: the settlement location cannot be said to be caused by chance.

**4** The NNI obtained depends very much on the area of the map in which the dots lie. Figure 14.7 shows the same distribution but on different-scale maps. The small-scale map gives an impression of clustering, while the large scale map suggests that the same points have a random distribution.

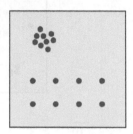

**Figure 14.6  Two types of point pattern on one map**

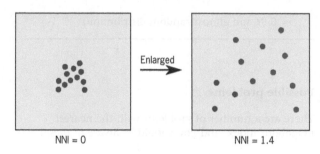

| NNI = 0 | Enlarged | NNI = 1.4 |

**Figure 14.7  A point pattern at two different scales**

*Linear nearest neighbour analysis*
If we wish to calculate a nearest neighbour index for points along a line, such as the distribution of shoe shops within a High Street or towns down the banks of a river, we can use a modified version of the nearest neighbour analysis. It can only be employed if the number of points is 10 or more.

**Method**

*Stage 1* Plot the points on a map.

*Stage 2* Measure the distance from each point to its nearest neighbouring point. Find the average of these distances $(\bar{D})$.

*Stage 3* Calculate the formula:

$$\text{NNI} = \frac{\bar{D}}{0.5 \left( \dfrac{L}{n-1} \right)}$$

where NNI = nearest neighbour index
$\bar{D}$ = mean observed nearest neighbour distance (stage 2)
$L$ = length of the line (such as the High Street)
$n$ = number of points (such as shops)

The result (NNI) is read off the scale:

| 0 | 1.0 | 2.0 |
|---|-----|-----|
| points clustered | points random | points uniform |

**Example**

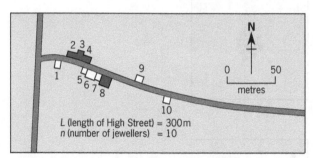

L (length of High Street) = 300 m
n (number of jewellers) = 10

**Figure 14.8  The location of jewellers in Bradford in 1896**

| Point | Nearest neighbour | Distance apart (m) |
|-------|-------------------|--------------------|
| 1 | 2 | 20 |
| 2 | 3 | 0 |
| 3 | 4 | 0 |
| 4 | 3 | 0 |
| 5 | 6 | 0 |
| 6 | 5 | 0 |
| 7 | 8 | 0 |
| 8 | 7 | 0 |
| 9 | 8 | 40 |
| 10 | 9 | 50 |
| | | 110 |

$\bar{D} = 110 \div 10 = 11$

$$\text{NNI} = \frac{11}{0.5 \left( \dfrac{300}{10-1} \right)} = 0.66 \text{ (moderately clustered)}$$

## 14.2 Index of dissimilarity

The index of dissimilarity is used to compare the distribution of two variables, especially different social or ethnic groups.

### Method

*Stage 1* Tabulate the data as shown. The data must be in percentage form and the total for each variable must be 100 per cent. Thus, in the example, the figure 53 per cent in column *x* does not mean that 53 per cent of the population of the South is black. It means that 53 per cent of black people in the USA live in the South.

*Stage 2* Find the difference between the two columns (always take the lower value from the higher, regardless of which column they are in). Add these differences and halve the result. This gives you the index of dissimilarity.

The possible range of values is between 0 and 100 per cent. It represents the proportion of group *x* that would have to move to a different area in order to have the same distribution as the group with which it is being compared. If measuring residential segregation of blacks and whites, 0 would mean that they had the same relative distribution, 100 would mean that no black person lived in the same area as any white person.

### Advantages

1 The index is simple to calculate.
2 The index has an actual verbal meaning, as described above.
3 The index is not affected by the relative sizes of the two groups being compared.
4 A large amount of information can be summarised very easily, as can be seen in Table 14.1.

### Example

Calculation of the index of dissimilarity (USA, 1980)

| Region | $x$<br>% of<br>black pop. | $y$<br>% of<br>white pop. | $\lvert x - y \rvert$ |
|---|---|---|---|
| South | 53 | 28 | 25 |
| North-East | 19 | 25 | 6 |
| North-Central | 20 | 29 | 9 |
| West | 8 | 18 | 10 |
| | 100 | 100 | $\Sigma 50$ |

$$\text{ID} = {}^1\!/_2 \sum \lvert x - y \rvert$$
$$\text{ID} = 50 \div 2$$
$$= 25$$

This result means that 25 per cent of the black population of the United States would have to change the region in which they live in order to have the same relative distribution as the white population, or vice versa.

### Limitation

The result is affected by the size of areal units within which the data is collected. The larger these units, the smaller the resulting index.

**Table 14.1 Index of dissimilarity for the residential distribution of selected ethnic groups in Chicago, USA, 1950 (*Source:* Duncan and Lieberson, 1959)**

| Country of origin | 1 | 2 | 3 | 4 | 5 | 6 | 7 | 8 | 9 | 10 | 11 |
|---|---|---|---|---|---|---|---|---|---|---|---|
| 1 England and Wales | – | 28.5 | 29.7 | 29.5 | 58.4 | 55.9 | 26.3 | 38.3 | 56.8 | 45.7 | 77.8 |
| 2 Ireland (Eire) | | – | 40.2 | 43.9 | 66.7 | 63.2 | 38.3 | 54.2 | 59.8 | 52.0 | 81.4 |
| 3 Sweden | | | – | 32.3 | 67.8 | 66.0 | 38.8 | 54.0 | 66.2 | 60.9 | 85.5 |
| 4 Germany | | | | – | 55.9 | 47.2 | 21.3 | 47.2 | 54.3 | 54.3 | 85.4 |
| 5 Poland | | | | | – | 43.5 | 47.3 | 58.3 | 50.8 | 52.6 | 90.8 |
| 6 Czechoslovakia | | | | | | – | 47.8 | 61.4 | 50.5 | 55.9 | 89.2 |
| 7 Austria | | | | | | | – | 45.6 | 53.8 | 45.6 | 82.5 |
| 8 USSR | | | | | | | | – | 67.5 | 57.5 | 87.1 |
| 9 Lithuania | | | | | | | | | – | 61.6 | 84.7 |
| 10 Italy | | | | | | | | | | – | 69.6 |
| 11 Black | | | | | | | | | | | – |

# 14.3 The Lorenz curve and Gini coefficient

*Lorenz curve*

The best use of the Lorenz curve is to show the degree of concentration of a variable in areas, such as the degree to which electronics manufacturing and coal mining in the UK is concentrated in certain regions.

## Method

*Stage 1* Lay out the data in a table as shown below.
*Stage 2* Divide the right-hand column by the left-hand column. This gives you the ratios between the two.
*Stage 3* Rank them according to these ratios and calculate the cumulative percentage for each of the two columns as shown below.
*Stage 4* Plot as a Lorenz curve as shown in Figure 14.9.

## Example

*Stage 1*

|  | A<br>Percentage of all UK jobs | B<br>Percentage of UK electronics jobs | C<br>B ÷ A | D<br>Rank of C |
|---|---|---|---|---|
| North | 5.1 | 4.2 | 0.82 | 6 |
| Yorkshire–Humberside | 8.8 | 3.6 | 0.41 | 11 |
| E. Midlands | 6.0 | 4.4 | 0.73 | 7 |
| W. Midlands | 9.1 | 4.1 | 0.45 | 10 |
| E. Anglia | 5.8 | 10.7 | 1.84 | 2 |
| South-East | 33.2 | 40.5 | 1.22 | 3 |
| South-West | 6.1 | 7.4 | 1.21 | 4 |
| North-West | 10.7 | 5.2 | 0.49 | 9 |
| Wales | 4.2 | 8.4 | 2.00 | 1 |
| Scotland | 9.0 | 10.1 | 1.12 | 5 |
| N. Ireland | 2.0 | 1.4 | 0.70 | 8 |
|  | 100 | 100 |  |  |

*Stage 2* Calculate the ratio between the percentage of jobs in electronics and the percentage of total employment: e.g. North = 4.2 divided by 5.1 = 0.82

*Stage 3* Rank them according to their ratios:

|  | E<br>Column C reordered in rank order | F<br>Cumulative % of all UK jobs | G<br>Cumulative % of UK electronics job |
|---|---|---|---|
| Wales | 2.00 | 4.2 | 8.4 |
| E. Anglia | 1.84 | 10.0 | 19.1 |
| South-East | 1.22 | 43.2 | 59.6 |
| South-West | 1.21 | 49.3 | 67.0 |
| Scotland | 1.12 | 58.3 | 77.1 |
| North | 0.82 | 63.4 | 81.3 |
| E. Midlands | 0.73 | 69.4 | 85.7 |
| N. Ireland | 0.70 | 71.4 | 87.1 |
| North-West | 0.49 | 82.1 | 92.3 |
| W. Midlands | 0.45 | 91.2 | 96.4 |
| Yorkshire–Humberside | 0.41 | 100.0 | 100.0 |

*Stage 4* See Figure 14.9

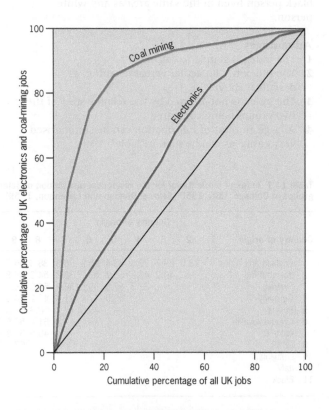

**Figure 14.9 Lorenz curve, showing the degree to which electronics and coal-mining jobs are concentrated in a limited number of regions in the UK**

*Gini coefficient*
The Gini coefficient is a means of expressing the Lorenz curve in numerical terms. It is found from the formula:

$$\text{Gini coefficient} = \frac{\text{Area of the graph above the diagonal representing even distribution}}{\text{Area of the graph between the diagonal and the curve}}$$

The area can be easily calculated by counting boxes and portions of bisected boxes on the graph paper. The Gini coefficient varies between 1 (the whole distribution concentrated in only one category) and infinity (totally even distribution). The same limitation as was made for Lorenz curves about the size of areal units in which data was collected and the comparability of two places applies to the Gini coefficient.

The Gini coefficient for electronics in Figure 14.9 is therefore as follows:

$$\text{Gini coefficient} = \frac{\text{Area above diagonal}}{\text{Area between diagonal and curve}}$$

$$= \frac{50}{38}$$

$$= 1.3$$

The Gini coefficient is useful because it is a more accurate indicator of concentration or diversity than the Lorenz curve. It is hard to visually compare two Lorenz curves that are only slightly different, but this difference will be brought out by comparing their respective Gini coefficients.

# 14.4 Location quotient

Like the Lorenz curve, the location quotient (LQ) is a measure of how concentrated a particular phenomenon is in one area: for example, how concentrated the jewellery industry of London is in Hatton Garden, or how concentrated coal mining in northern England is in County Durham. To calculate the location quotient, we need to know the distribution of the phenomenon under study (the jewellery industry) and the distribution of the wider population of which the phenomenon is just a part (all industry). The calculation is simple:

| Percentage of London's jewellery industry in Hatton Garden (in terms of value of output) | Percentage of all London's industry in Hatton Garden | Location quotient |
|---|---|---|
| 13 | 0.002 | (13 ÷ 0.002) = 6500 |

The range of possible values is from zero to infinity. If the LQ is greater than 1.0, then the sub-group (jewellery industry) is over-represented. The larger the LQ, the greater the concentration of the sub-group.

# 15 Presentation and layout

## 15.1 Project structure

An A-level individual study is not marked according to the result you obtained (you do not lose marks if your hypothesis is disproved) but according to how you set about collecting data, analysing them and writing up your findings, i.e. your method.

Projects are expected to be researched and presented in a scientific manner, and the format suggested below is what the majority of A-level examiners will be looking for. By following this pattern you are less likely to fall into the trap of presenting a purely descriptive project and more likely to adopt the hypothesis-testing or problem-solving or question-answering method favoured by most examination boards. Note that some boards require a slightly different project structure: you must check.

### Preliminary pages
1 Project title, your name and candidate number, school and centre number.
2 List of contents with page numbers.
3 List of diagrams and illustrations with page numbers.

### Chapter 1

#### Aim
Describe the aim of your project. This may be to test a hypothesis or theory, in which case set it out. Justify your hypotheses. You should have sound geographical reasons for believing the hypotheses to be correct. It is a good idea to say what sorts of results you expect to find. This will involve referring to past research on the topic. You will compare your own aim with relevant theory or earlier research on the same topic.

#### Introduction
This sets the scene for the project and includes a description of the study area, including maps at different scales, and, where relevant, a brief history of the study area (e.g. for urban/population studies).

### Chapter 2

#### Techniques
Describe all the techniques and data sources you have employed, and state clearly what the techniques were used for. Include the *dates* and *times* the data were collected and explain why these dates and times were used. Evaluate and criticise techniques and data sources.

What problems did you encounter? What are the likely sources of error? What bias was there and what steps did you take to overcome it?

The techniques chapter must contain a section on sampling. How did you choose the exact point at which you took your sample? Was your sample stratified, linear, random or whatever? How many samples (temperature readings, interviews, etc.) did you take? How did you arrive at this number? In ideal conditions, how many samples would you have liked to take?

### Chapter 3

#### Results and analysis
Going through hypothesis by hypothesis and using a wide variety of graphs, you should describe and then explain the results of your investigation. State the broad trends, average figures and some extremes (maxima and minima). All graphs should carry a title and a figure number, and should be referred to in the text. All graphs and tables should carry a small footnote giving the source of the information (for example: questionnaire survey, 1991 census, etc.). Point out any unusual or unexpected results.

The results chapter will contain descriptive statistics: range, interquartile range, mean, median, mode (as appropriate), standard deviation, dispersion graphs, etc. You may well also include some analytical statistics such as correlations and tests of significance.

Explain your results. Are they what you expected them to be? Do the results match your hypotheses/theories or are they slightly or very different? If the results agree with your expectations, you must now analyse them and explain the reasons for this pattern.

If your results do not agree with your expectations, you must give explanations for why they do not. Were your readings faulty? Was your sample too small? Were your expectations wrong? Was your theory flawed in some way? What characteristics does your subject have that explain why it does not match the theory?

For many projects it is appropriate, having presented the results, to consider the policy implications for national and local planning authorities. If problems, such as land-use conflicts, have emerged, what can or should be done to resolve them?

Issue-based enquiries should include some element of judgement. This normally requires a consideration of the purpose of the project from the perspective of different groups of people. You may be expected to express your own opinion based upon the results of your research.

*Chapter 4*

### Conclusion

Your conclusion should recall and summarise the main findings and may comment briefly on their wider significance. There could usefully be a brief section on what you have learnt from the exercise and how you might have improved upon it. What further research could be done to follow up? Remember, a good project will raise as many questions as it answers.

### Final pages

APPENDICES containing tables, selected raw data, etc.

BIBLIOGRAPHY which should contain, for each source you have used, the author's name, book/journal title, publisher and date of publication.

ACKNOWLEDGEMENTS of assistance given.

## 15.2 General points to observe

1  Keep your field notebook and all your data. They may be needed at your interview if your project is externally moderated.
2  Include some photographs if possible, but not too many. They should all be referred to in the text. Always ask yourself: 'Does this picture serve a purpose?'
3  It is a good idea to type your project on a computer. This makes it very much easier for you to alter what you have written and add or delete. As you type your project on the computer, keep saving at regular, frequent intervals and keep a backup copy. Print on one side of the paper only, using double spacing; keep a 3 cm left-hand margin.
4  Use subheadings and clearly 'signpost' your project so that it is obvious to the examiner what academic path you are following.
5  Stick to the prescribed word length laid down by your board.
6  Use digital mapping software to create maps if possible and ensure all maps have a scale and compass arrow.

# Appendices

## A1 Random sampling numbers

| | | | | | | | | | | | | | | | | | |
|---|---|---|---|---|---|---|---|---|---|---|---|---|---|---|---|---|---|
| 20 | 17 | 42 | 28 | 23 | 17 | 59 | 66 | 38 | 61 | 02 | 10 | 86 | 10 | 51 | 55 | 92 | 52 |
| 74 | 49 | 04 | 49 | 03 | 04 | 10 | 33 | 53 | 70 | 11 | 54 | 48 | 63 | 94 | 60 | 94 | 49 |
| 94 | 70 | 49 | 31 | 38 | 67 | 23 | 42 | 29 | 65 | 40 | 88 | 78 | 71 | 37 | 18 | 48 | 64 |
| 22 | 15 | 78 | 15 | 69 | 84 | 32 | 52 | 32 | 54 | 15 | 12 | 54 | 02 | 01 | 37 | 38 | 37 |
| 93 | 29 | 12 | 18 | 27 | 30 | 30 | 55 | 91 | 87 | 50 | 57 | 58 | 51 | 49 | 36 | 12 | 53 |
| 45 | 04 | 77 | 97 | 36 | 14 | 99 | 45 | 52 | 95 | 69 | 85 | 03 | 83 | 51 | 87 | 85 | 56 |
| 44 | 91 | 99 | 49 | 89 | 39 | 94 | 60 | 48 | 49 | 06 | 77 | 64 | 72 | 59 | 26 | 08 | 51 |
| 16 | 23 | 91 | 02 | 19 | 96 | 47 | 59 | 89 | 65 | 27 | 84 | 30 | 92 | 63 | 37 | 26 | 24 |
| 04 | 50 | 65 | 04 | 65 | 65 | 82 | 42 | 70 | 51 | 55 | 04 | 61 | 47 | 88 | 83 | 99 | 34 |
| 32 | 70 | 17 | 72 | 03 | 61 | 66 | 26 | 24 | 71 | 22 | 77 | 88 | 33 | 17 | 78 | 08 | 92 |
| 03 | 64 | 59 | 07 | 42 | 95 | 81 | 39 | 06 | 41 | 20 | 81 | 92 | 34 | 51 | 90 | 39 | 08 |
| 62 | 49 | 00 | 90 | 67 | 86 | 83 | 48 | 31 | 83 | 19 | 07 | 67 | 68 | 49 | 03 | 27 | 47 |
| 61 | 00 | 95 | 86 | 98 | 36 | 14 | 03 | 48 | 88 | 51 | 07 | 33 | 40 | 06 | 86 | 33 | 76 |
| 89 | 03 | 90 | 49 | 28 | 74 | 21 | 04 | 09 | 96 | 60 | 45 | 22 | 03 | 52 | 80 | 01 | 79 |
| 01 | 72 | 33 | 85 | 52 | 40 | 60 | 07 | 06 | 71 | 89 | 27 | 14 | 29 | 55 | 24 | 85 | 79 |
| 27 | 56 | 49 | 79 | 34 | 34 | 32 | 22 | 60 | 53 | 91 | 17 | 33 | 26 | 44 | 70 | 93 | 14 |
| 49 | 05 | 74 | 48 | 10 | 55 | 35 | 25 | 24 | 28 | 20 | 22 | 35 | 66 | 66 | 34 | 26 | 35 |
| 49 | 74 | 37 | 25 | 97 | 26 | 33 | 94 | 42 | 23 | 01 | 28 | 59 | 58 | 92 | 69 | 03 | 66 |
| 20 | 26 | 22 | 43 | 88 | 08 | 19 | 85 | 08 | 12 | 47 | 65 | 65 | 63 | 56 | 07 | 97 | 85 |
| 48 | 87 | 77 | 96 | 43 | 39 | 76 | 93 | 08 | 79 | 22 | 18 | 54 | 55 | 93 | 75 | 97 | 26 |
| 08 | 72 | 87 | 46 | 75 | 73 | 00 | 11 | 27 | 07 | 05 | 20 | 30 | 85 | 22 | 21 | 04 | 67 |
| 95 | 97 | 98 | 62 | 17 | 27 | 31 | 42 | 64 | 71 | 46 | 22 | 32 | 75 | 19 | 32 | 20 | 99 |
| 37 | 99 | 57 | 31 | 70 | 40 | 46 | 55 | 46 | 12 | 24 | 32 | 36 | 74 | 69 | 20 | 72 | 10 |
| 05 | 79 | 58 | 37 | 85 | 33 | 75 | 18 | 88 | 71 | 23 | 44 | 54 | 28 | 00 | 48 | 96 | 23 |
| 55 | 85 | 63 | 42 | 00 | 79 | 91 | 22 | 29 | 01 | 41 | 39 | 51 | 50 | 36 | 65 | 26 | 11 |
| 67 | 28 | 96 | 25 | 68 | 36 | 24 | 72 | 03 | 85 | 49 | 24 | 05 | 69 | 64 | 86 | 08 | 19 |
| 85 | 86 | 94 | 78 | 32 | 59 | 51 | 82 | 86 | 43 | 73 | 84 | 45 | 60 | 89 | 57 | 06 | 87 |
| 40 | 10 | 60 | 09 | 05 | 88 | 78 | 44 | 63 | 13 | 58 | 25 | 37 | 11 | 18 | 47 | 75 | 62 |
| 94 | 55 | 89 | 48 | 90 | 80 | 77 | 80 | 26 | 89 | 87 | 44 | 23 | 74 | 66 | 20 | 20 | 19 |
| 11 | 63 | 77 | 77 | 23 | 20 | 33 | 62 | 62 | 19 | 29 | 03 | 94 | 15 | 56 | 37 | 14 | 09 |
| 64 | 00 | 26 | 04 | 54 | 55 | 38 | 57 | 94 | 62 | 68 | 40 | 26 | 04 | 24 | 25 | 03 | 61 |
| 50 | 94 | 13 | 23 | 78 | 41 | 60 | 58 | 10 | 60 | 88 | 46 | 30 | 21 | 45 | 98 | 70 | 96 |
| 66 | 98 | 37 | 96 | 44 | 13 | 45 | 05 | 34 | 59 | 75 | 85 | 48 | 97 | 27 | 19 | 17 | 85 |
| 66 | 91 | 42 | 83 | 60 | 77 | 90 | 91 | 60 | 90 | 79 | 62 | 57 | 66 | 72 | 28 | 08 | 70 |
| 33 | 58 | 12 | 18 | 02 | 07 | 19 | 40 | 21 | 29 | 39 | 45 | 90 | 42 | 58 | 84 | 85 | 43 |
| 52 | 49 | 70 | 16 | 72 | 40 | 73 | 05 | 50 | 90 | 02 | 04 | 98 | 24 | 05 | 30 | 27 | 25 |
| 74 | 98 | 93 | 99 | 78 | 30 | 79 | 47 | 96 | 62 | 45 | 58 | 40 | 37 | 89 | 76 | 84 | 41 |
| 50 | 26 | 54 | 30 | 01 | 88 | 69 | 57 | 54 | 45 | 69 | 88 | 23 | 21 | 05 | 69 | 93 | 44 |
| 49 | 46 | 61 | 89 | 33 | 79 | 96 | 84 | 28 | 34 | 19 | 35 | 28 | 73 | 39 | 59 | 56 | 34 |
| 19 | 64 | 13 | 44 | 78 | 39 | 73 | 88 | 62 | 03 | 36 | 00 | 25 | 96 | 86 | 76 | 67 | 90 |
| 64 | 17 | 47 | 67 | 87 | 59 | 81 | 40 | 72 | 61 | 14 | 00 | 28 | 28 | 55 | 86 | 23 | 38 |
| 18 | 43 | 97 | 37 | 68 | 97 | 56 | 56 | 57 | 95 | 01 | 88 | 11 | 89 | 48 | 07 | 42 | 07 |
| 65 | 58 | 60 | 87 | 51 | 09 | 96 | 61 | 15 | 53 | 66 | 81 | 66 | 88 | 44 | 75 | 37 | 01 |
| 79 | 90 | 31 | 00 | 91 | 14 | 85 | 65 | 31 | 75 | 43 | 15 | 45 | 93 | 64 | 78 | 34 | 53 |
| 07 | 23 | 00 | 15 | 59 | 05 | 16 | 09 | 94 | 42 | 20 | 40 | 63 | 76 | 65 | 67 | 34 | 11 |
| 90 | 98 | 14 | 24 | 01 | 51 | 95 | 46 | 30 | 32 | 33 | 19 | 00 | 14 | 19 | 28 | 40 | 51 |
| 53 | 82 | 62 | 02 | 21 | 82 | 34 | 13 | 41 | 03 | 12 | 85 | 65 | 30 | 00 | 97 | 56 | 30 |
| 98 | 17 | 26 | 15 | 04 | 50 | 76 | 25 | 20 | 33 | 54 | 84 | 39 | 31 | 23 | 33 | 59 | 64 |
| 08 | 91 | 12 | 44 | 82 | 40 | 30 | 62 | 45 | 50 | 64 | 54 | 65 | 17 | 89 | 25 | 59 | 44 |
| 37 | 21 | 46 | 77 | 84 | 87 | 67 | 39 | 85 | 54 | 97 | 37 | 33 | 41 | 11 | 74 | 90 | 50 |

After Lindley and Miller (1953)

## Rules

1  You can start reading from any point and move in any direction *as long as you are consistent*, i.e. if you start top left and read along rows, you must continue along the rows.

2  Numbers can be read singly, in pairs (as printed),or multiples of 3, 4, etc. Thus in the first row you can read 2, or 20, or 201, etc. Decide which you want and *be consistent*.

# A2 Table of z-scores

| z | Column A $p$ | Column B $p_1$ | Column C $p_2$ |
|---|---|---|---|
| 0.0 | 0.000 | 0.500 | 1.000 |
| 0.1 | 0.040 | 0.460 | 0.920 |
| 0.2 | 0.079 | 0.421 | 0.841 |
| 0.3 | 0.118 | 0.382 | 0.764 |
| 0.4 | 0.155 | 0.345 | 0.689 |
| 0.5 | 0.191 | 0.309 | 0.617 |
| 0.6 | 0.226 | 0.274 | 0.549 |
| 0.7 | 0.258 | 0.242 | 0.484 |
| 0.8 | 0.288 | 0.212 | 0.424 |
| 0.9 | 0.316 | 0.184 | 0.368 |
| 1.0 | 0.341 | 0.159 | 0.317 |
| 1.1 | 0.364 | 0.136 | 0.271 |
| 1.2 | 0.385 | 0.115 | 0.230 |
| 1.3 | 0.403 | 0.097 | 0.193 |
| 1.4 | 0.419 | 0.081 | 0.162 |
| 1.5 | 0.433 | 0.067 | 0.134 |
| 1.6 | 0.445 | 0.055 | 0.110 |
| 1.7 | 0.455 | 0.045 | 0.089 |
| 1.8 | 0.464 | 0.036 | 0.072 |
| 1.9 | 0.471 | 0.029 | 0.057 |
| 1.96* | 0.475 | 0.025 | 0.050 |
| 2.0 | 0.477 | 0.023 | 0.046 |
| 2.1 | 0.482 | 0.018 | 0.036 |
| 2.2 | 0.486 | 0.014 | 0.028 |
| 2.3 | 0.489 | 0.011 | 0.021 |
| 2.4 | 0.492 | 0.008 | 0.016 |
| 2,5 | 0.494 | 0.006 | 0.012 |
| 2.28* | 0.495 | 0.005 | 0.010 |
| 2.6 | 0.495 | 0.005 | 0.009 |
| 2.7 | 0.496 | 0.004 | 0.007 |
| 2.8 | 0.497 | 0.003 | 0.005 |
| 2.9 | 0.498 | 0.002 | 0.004 |
| 3.0 | 0.499 | 0.001 | 0.003 |
| 3.1 | 0.499 | 0.001 | 0.002 |
| 3.2 | 0.499 | 0.001 | 0.001 |
| 3.3 | 0.499 | 0.001 | 0.001 |
| 3.4 | 0.500 | 0.000 | 0.001 |
| 3.5 | 0.500 | 0.000 | 0.000 |

# A3 Student's t-tables

| Degrees of freedom | Rejection level probabilities | | | | |
|---|---|---|---|---|---|
| | $p = 0.1$ | $p = 0.05$ | $p = 0.02$ | $p = 0.01$ | $p = 0.001$ |
| 1 | 6.31 | 12.71 | 31.82 | 63.66 | 636.62 |
| 2 | 2.92 | 4.30 | 6.97 | 9.93 | 31.60 |
| 3 | 2.35 | 3.18 | 4.54 | 5.84 | 12.94 |
| 4 | 2.13 | 2.78 | 3.75 | 4.60 | 8.61 |
| 5 | 2.02 | 2.57 | 3.37 | 4.03 | 6.86 |
| 6 | 1.94 | 2.45 | 3.14 | 3.71 | 5.96 |
| 7 | 1.90 | 2.37 | 3.00 | 3.50 | 5.41 |
| 8 | 1.86 | 2.31 | 2.90 | 3.36 | 5.04 |
| 9 | 1.83 | 2.26 | 2.82 | 3.25 | 4.78 |
| 10 | 1.81 | 2.23 | 2.76 | 3.17 | 4.59 |
| 11 | 1.80 | 2.20 | 2.75 | 3.11 | 4.44 |
| 12 | 1.78 | 2.18 | 2.68 | 3.06 | 4.32 |
| 13 | 1.77 | 2.16 | 2.65 | 3.01 | 4.22 |
| 14 | 1.76 | 2.15 | 2.62 | 2.98 | 4.14 |
| 15 | 1.75 | 2.13 | 2.60 | 2.95 | 4.07 |
| 16 | 1.75 | 2.12 | 2.58 | 2.92 | 4.02 |
| 17 | 1.74 | 2.11 | 2.57 | 2.90 | 3.97 |
| 18 | 1.73 | 2.10 | 2.55 | 2.88 | 3.92 |
| 19 | 1.73 | 2.09 | 2.54 | 2.86 | 3.88 |
| 20 | 1.73 | 2.09 | 2.53 | 2.85 | 3.85 |
| 21 | 1.72 | 2.08 | 2.52 | 2.83 | 3.82 |
| 22 | 1.72 | 2.07 | 2.51 | 2.82 | 3.79 |
| 23 | 1.71 | 2.07 | 2.50 | 2.81 | 3.77 |
| 24 | 1.71 | 2.06 | 2.49 | 2.80 | 3.75 |
| 25 | 1.71 | 2.06 | 2.49 | 2.79 | 3.73 |
| 26 | 1.71 | 2.06 | 2.48 | 2.78 | 3.71 |
| 27 | 1.70 | 2.05 | 2.47 | 2.77 | 3.69 |
| 28 | 1.70 | 2.05 | 2.47 | 2.76 | 3.67 |
| 29 | 1.70 | 2.05 | 2.46 | 2.76 | 3.66 |
| 30 | 1.70 | 2.04 | 2.46 | 2.75 | 3.65 |
| 40 | 1.68 | 2.02 | 2.42 | 2.70 | 3.55 |
| 60 | 1.67 | 2.00 | 2.00 | 2.66 | 3.46 |

Note:
Col. A: $p$ = the probability of a value lying between the mean and the corresponding value of z.
Col. B: $p_1$ = the probability of a value exceeding the given value of z (a one-tailed probability).
Col. C: $p_2$ = the probability of a value exceeding either $+z$ or $-z$ (a two-tailed probability).

*Critical values of z corresponding to the 0.05 and 0.01 levels (two-tailed) have been given to two decimal places.

After Lindley and Miller (1953)

# A4 Critical values of Mann–Whitney $U$

Tables A–F: the probability read off the table must be less than the rejection level if $U$ is significant.
Tables G–J: the calculated value of $U$ must be less than the critical value if $U$ is to be regarded as significant.

## A

**$n_2 = 3$**

| $U$ \ $n_1$ | 1 | 2 | 3 |
|---|---|---|---|
| 0 | 0.250 | 0.100 | 0.050 |
| 1 | 0.500 | 0.200 | 0.100 |

## B

**$n_2 = 4$**

| $U$ \ $n_1$ | 1 | 2 | 3 | 4 |
|---|---|---|---|---|
| 0 | 0.200 | 0.067 | 0.028 | 0.014 |
| 1 | 0.400 | 0.133 | 0.057 | 0.029 |
| 2 | 0.600 | 0.267 | 0.114 | 0.057 |
| 3 |  | 0.400 | 0.200 | 0.100 |

## C

**$n_2 = 5$**

| $U$ \ $n_1$ | 1 | 2 | 3 | 4 | 5 |
|---|---|---|---|---|---|
| 0 | 0.167 | 0.047 | 0.018 | 0.008 | 0.004 |
| 1 | 0.333 | 0.095 | 0.036 | 0.016 | 0.008 |
| 2 | 0.500 | 0.190 | 0.071 | 0.032 | 0.016 |
| 3 | 0.667 | 0.286 | 0.125 | 0.056 | 0.028 |
| 4 |  | 0.429 | 0.196 | 0.095 | 0.048 |
|  |  | 0.571 | 0.286 | 0.143 | 0.075 |

## D

**$n_2 = 6$**

| $U$ \ $n_1$ | 1 | 2 | 3 | 4 | 5 | 6 |
|---|---|---|---|---|---|---|
| 0 | 0.143 | 0.036 | 0.012 | 0.005 | 0.002 | 0.001 |
| 1 | 0.286 | 0.071 | 0.024 | 0.010 | 0.004 | 0.002 |
| 2 | 0.428 | 0.143 | 0.048 | 0.019 | 0.009 | 0.004 |
| 3 | 0.571 | 0.214 | 0.183 | 0.033 | 0.015 | 0.008 |
| 4 |  | 0.321 | 0.131 | 0.057 | 0.026 | 0.013 |
| 5 |  | 0.429 | 0.190 | 0.086 | 0.041 | 0.021 |
| 6 |  | 0.571 | 0.274 | 0.129 | 0.063 | 0.032 |
| 7 |  |  | 0.357 | 0.176 | 0.089 | 0.047 |
| 8 |  |  | 0.452 | 0.238 | 0.123 | 0.066 |
| 9 |  |  | 0.548 | 0.305 | 0.165 | 0.090 |

## E

**$n_2 = 7$**

| $U$ \ $n_1$ | 1 | 2 | 3 | 4 | 5 | 6 | 7 |
|---|---|---|---|---|---|---|---|
| 0 | 0.125 | 0.028 | 0.008 | 0.003 | 0.001 | 0.001 | 0.000 |
| 1 | 0.250 | 0.056 | 0.017 | 0.006 | 0.003 | 0.001 | 0.001 |
| 2 | 0.375 | 0.111 | 0.033 | 0.012 | 0.005 | 0.002 | 0.001 |
| 3 | 0.500 | 0.167 | 0.058 | 0.021 | 0.009 | 0.004 | 0.002 |
| 4 | 0.625 | 0.250 | 0.092 | 0.036 | 0.015 | 0.007 | 0.003 |
| 5 |  | 0.333 | 0.133 | 0.055 | 0.024 | 0.011 | 0.006 |
| 6 |  | 0.444 | 0.192 | 0.082 | 0.037 | 0.017 | 0.009 |
| 7 |  | 0.556 | 0.258 | 0.115 | 0.053 | 0.026 | 0.013 |
| 8 |  |  | 0.333 | 0.158 | 0.074 | 0.037 | 0.019 |
| 9 |  |  | 0.417 | 0.206 | 0.101 | 0.051 | 0.027 |
| 10 |  |  | 0.500 | 0.264 | 0.134 | 0.069 | 0.036 |
| 11 |  |  | 0.583 | 0.324 | 0.172 | 0.090 | 0.049 |
| 12 |  |  |  | 0.394 | 0.216 | 0.117 | 0.064 |
| 13 |  |  |  | 0.464 | 0.265 | 0.147 | 0.082 |

## F

**$n_2 = 8$**

| $U$ \ $n_1$ | 1 | 2 | 3 | 4 | 5 | 6 | 7 | 8 |
|---|---|---|---|---|---|---|---|---|
| 0 | 0.111 | 0.022 | 0.006 | 0.002 | 0.001 | 0.000 | 0.000 | 0.000 |
| 1 | 0.222 | 0.044 | 0.012 | 0.004 | 0.002 | 0.001 | 0.000 | 0.000 |
| 2 | 0.333 | 0.089 | 0.024 | 0.008 | 0.003 | 0.001 | 0.001 | 0.000 |
| 3 | 0.444 | 0.133 | 0.042 | 0.014 | 0.005 | 0.002 | 0.001 | 0.001 |
| 4 | 0.556 | 0.200 | 0.067 | 0.024 | 0.009 | 0.004 | 0.002 | 0.001 |
| 5 |  | 0.267 | 0.097 | 0.036 | 0.015 | 0.006 | 0.003 | 0.001 |
| 6 |  | 0.356 | 0.139 | 0.055 | 0.023 | 0.010 | 0.005 | 0.002 |
| 7 |  | 0.444 | 0.188 | 0.077 | 0.033 | 0.015 | 0.007 | 0.003 |
| 8 |  | 0.556 | 0.248 | 0.107 | 0.047 | 0.021 | 0.010 | 0.005 |
| 9 |  |  | 0.315 | 0.141 | 0.064 | 0.030 | 0.014 | 0.007 |
| 10 |  |  | 0.387 | 0.184 | 0.085 | 0.041 | 0.020 | 0.010 |
| 11 |  |  | 0.461 | 0.230 | 0.111 | 0.054 | 0.027 | 0.014 |
| 12 |  |  | 0.539 | 0.285 | 0.142 | 0.071 | 0.036 | 0.019 |
| 13 |  |  |  | 0.341 | 0.177 | 0.091 | 0.047 | 0.025 |
| 14 |  |  |  | 0.404 | 0.217 | 0.114 | 0.060 | 0.032 |
| 15 |  |  |  | 0.467 | 0.262 | 0.141 | 0.076 | 0.041 |
| 16 |  |  |  | 0.533 | 0.311 | 0.172 | 0.095 | 0.052 |
| 17 |  |  |  |  | 0.362 | 0.207 | 0.116 | 0.065 |
| 18 |  |  |  |  | 0.416 | 0.245 | 0.140 | 0.080 |
| 19 |  |  |  |  | 0.472 | 0.286 | 0.168 | 0.097 |

From Siegal (1956); after Mann and Whitney (1947).

## G

**Critical values of U for a two-tailed test at α = 0.002 and a one-tailed test at α = 0.001**

| $n_1$ \ $n_2$ | 9 | 10 | 11 | 12 | 13 | 14 | 15 | 16 | 17 | 18 | 19 | 20 |
|---|---|---|---|---|---|---|---|---|---|---|---|---|
| 1 |  |  |  |  |  |  |  |  |  |  |  |  |
| 2 |  |  |  |  |  |  |  |  |  |  |  |  |
| 3 |  |  |  |  |  |  |  |  | 0 | 0 | 0 | 0 |
| 4 |  | 0 | 0 | 0 | 1 | 1 | 1 | 2 | 2 | 3 | 3 | 3 |
| 5 | 1 | 1 | 2 | 2 | 3 | 3 | 4 | 5 | 5 | 6 | 7 | 7 |
| 6 | 2 | 3 | 4 | 4 | 5 | 6 | 7 | 8 | 9 | 10 | 11 | 12 |
| 7 | 3 | 5 | 6 | 7 | 8 | 9 | 10 | 11 | 13 | 14 | 15 | 16 |
| 8 | 5 | 6 | 8 | 9 | 11 | 12 | 14 | 15 | 17 | 18 | 20 | 21 |
| 9 | 7 | 8 | 10 | 12 | 14 | 15 | 17 | 19 | 21 | 23 | 25 | 26 |
| 10 | 8 | 10 | 12 | 14 | 17 | 19 | 21 | 23 | 25 | 27 | 29 | 32 |
| 11 | 10 | 12 | 15 | 17 | 20 | 22 | 24 | 27 | 29 | 32 | 34 | 37 |
| 12 | 12 | 14 | 17 | 20 | 23 | 25 | 28 | 31 | 34 | 37 | 40 | 42 |
| 13 | 14 | 17 | 20 | 23 | 26 | 29 | 32 | 35 | 38 | 42 | 45 | 48 |
| 14 | 15 | 19 | 22 | 25 | 29 | 32 | 36 | 39 | 43 | 46 | 50 | 54 |
| 15 | 17 | 21 | 24 | 28 | 32 | 36 | 40 | 43 | 47 | 51 | 55 | 59 |
| 16 | 19 | 23 | 27 | 31 | 35 | 39 | 43 | 48 | 52 | 56 | 60 | 65 |
| 17 | 21 | 25 | 29 | 34 | 38 | 43 | 47 | 52 | 57 | 61 | 66 | 70 |
| 18 | 23 | 27 | 32 | 37 | 42 | 46 | 51 | 56 | 61 | 66 | 71 | 76 |
| 19 | 25 | 29 | 34 | 40 | 45 | 50 | 55 | 60 | 66 | 71 | 77 | 82 |
| 20 | 26 | 32 | 37 | 42 | 48 | 54 | 59 | 65 | 70 | 76 | 82 | 88 |

## H

**Critical values of U for a two-tailed test at α = 0.02 and a one-tailed test at α = 0.01**

| $n_1$ \ $n_2$ | 9 | 10 | 11 | 12 | 13 | 14 | 15 | 16 | 17 | 18 | 19 | 20 |
|---|---|---|---|---|---|---|---|---|---|---|---|---|
| 1 |  |  |  |  |  |  |  |  |  |  |  |  |
| 2 |  |  |  |  | 0 | 0 | 0 | 0 | 0 | 0 | 1 | 1 |
| 3 | 1 | 1 | 1 | 2 | 2 | 2 | 3 | 3 | 4 | 4 | 4 | 5 |
| 4 | 3 | 3 | 4 | 5 | 5 | 6 | 7 | 7 | 8 | 9 | 9 | 10 |
| 5 | 5 | 6 | 7 | 8 | 9 | 10 | 11 | 12 | 13 | 14 | 15 | 16 |
| 6 | 7 | 8 | 9 | 11 | 12 | 13 | 15 | 16 | 18 | 19 | 20 | 22 |
| 7 | 9 | 11 | 12 | 14 | 16 | 17 | 19 | 21 | 23 | 24 | 26 | 28 |
| 8 | 11 | 13 | 15 | 17 | 20 | 22 | 24 | 26 | 28 | 30 | 32 | 34 |
| 9 | 14 | 16 | 18 | 21 | 23 | 26 | 28 | 31 | 33 | 36 | 38 | 40 |
| 10 | 16 | 19 | 22 | 24 | 27 | 30 | 33 | 36 | 38 | 41 | 44 | 47 |
| 11 | 18 | 22 | 25 | 28 | 31 | 34 | 37 | 41 | 44 | 47 | 50 | 53 |
| 12 | 21 | 24 | 28 | 31 | 35 | 38 | 42 | 46 | 49 | 53 | 56 | 60 |
| 13 | 23 | 27 | 31 | 35 | 39 | 43 | 47 | 51 | 55 | 59 | 63 | 67 |
| 14 | 26 | 30 | 34 | 38 | 43 | 47 | 51 | 56 | 60 | 65 | 69 | 73 |
| 15 | 28 | 33 | 37 | 42 | 47 | 51 | 56 | 61 | 66 | 70 | 75 | 80 |
| 16 | 31 | 36 | 41 | 46 | 51 | 56 | 61 | 66 | 71 | 76 | 82 | 87 |
| 17 | 33 | 38 | 44 | 49 | 55 | 60 | 66 | 71 | 77 | 82 | 88 | 93 |
| 18 | 36 | 41 | 47 | 53 | 59 | 65 | 70 | 76 | 82 | 88 | 94 | 100 |
| 19 | 38 | 44 | 50 | 56 | 63 | 69 | 75 | 82 | 88 | 94 | 101 | 107 |
| 20 | 40 | 47 | 53 | 60 | 67 | 73 | 80 | 87 | 93 | 100 | 107 | 114 |

## I

**Critical values of U for a two-tailed test at α = 0.05 and a one-tailed test at α = 0.025**

| $n_1$ \ $n_2$ | 9 | 10 | 11 | 12 | 13 | 14 | 15 | 16 | 17 | 18 | 19 | 20 |
|---|---|---|---|---|---|---|---|---|---|---|---|---|
| 1 |  |  |  |  |  |  |  |  |  |  |  |  |
| 2 | 0 | 0 | 0 | 1 | 1 | 1 | 1 | 1 | 2 | 2 | 2 | 2 |
| 3 | 2 | 3 | 3 | 4 | 4 | 5 | 5 | 6 | 6 | 7 | 7 | 8 |
| 4 | 4 | 5 | 6 | 7 | 8 | 9 | 10 | 11 | 11 | 12 | 13 | 13 |
| 5 | 7 | 8 | 9 | 11 | 12 | 13 | 14 | 15 | 17 | 18 | 19 | 20 |
| 6 | 10 | 11 | 13 | 14 | 16 | 17 | 19 | 21 | 22 | 24 | 25 | 27 |
| 7 | 12 | 14 | 16 | 18 | 20 | 22 | 24 | 26 | 28 | 30 | 32 | 34 |
| 8 | 15 | 17 | 19 | 22 | 24 | 26 | 29 | 31 | 34 | 36 | 38 | 41 |
| 9 | 17 | 20 | 23 | 26 | 28 | 31 | 34 | 37 | 39 | 42 | 45 | 48 |
| 10 | 20 | 23 | 26 | 29 | 33 | 36 | 39 | 42 | 45 | 48 | 52 | 55 |
| 11 | 23 | 26 | 30 | 33 | 37 | 40 | 44 | 47 | 51 | 55 | 58 | 62 |
| 12 | 26 | 29 | 33 | 37 | 41 | 45 | 49 | 53 | 57 | 61 | 65 | 69 |
| 13 | 28 | 33 | 37 | 41 | 45 | 50 | 54 | 59 | 63 | 67 | 72 | 76 |
| 14 | 31 | 36 | 40 | 45 | 50 | 55 | 59 | 64 | 67 | 74 | 78 | 83 |
| 15 | 34 | 39 | 44 | 49 | 54 | 59 | 64 | 70 | 75 | 80 | 85 | 90 |
| 16 | 37 | 42 | 47 | 53 | 59 | 64 | 70 | 75 | 81 | 86 | 92 | 98 |
| 17 | 39 | 45 | 51 | 57 | 63 | 67 | 75 | 81 | 87 | 93 | 99 | 105 |
| 18 | 42 | 48 | 55 | 61 | 67 | 74 | 80 | 86 | 93 | 99 | 106 | 112 |
| 19 | 45 | 52 | 58 | 65 | 72 | 78 | 85 | 92 | 99 | 106 | 113 | 119 |
| 20 | 48 | 55 | 62 | 69 | 76 | 83 | 90 | 98 | 105 | 112 | 119 | 127 |

From Siegal (1956); after Auble (1953).

## J

**Critical values of U for a two-tailed test at α = 0.10 and a one-tailed test at α = 0.05**

| $n_1$ \ $n_2$ | 9 | 10 | 11 | 12 | 13 | 14 | 15 | 16 | 17 | 18 | 19 | 20 |
|---|---|---|---|---|---|---|---|---|---|---|---|---|
| 1 |  |  |  |  |  |  |  |  |  |  | 0 | 0 |
| 2 | 1 | 1 | 1 | 2 | 2 | 2 | 3 | 3 | 3 | 4 | 4 | 4 |
| 3 | 3 | 4 | 5 | 5 | 6 | 7 | 7 | 8 | 9 | 9 | 10 | 11 |
| 4 | 6 | 7 | 8 | 9 | 10 | 11 | 12 | 14 | 15 | 16 | 17 | 18 |
| 5 | 9 | 11 | 12 | 13 | 15 | 16 | 18 | 19 | 20 | 22 | 23 | 25 |
| 6 | 12 | 14 | 16 | 17 | 19 | 21 | 23 | 25 | 26 | 28 | 30 | 32 |
| 7 | 15 | 17 | 19 | 21 | 24 | 26 | 28 | 30 | 33 | 35 | 37 | 39 |
| 8 | 18 | 20 | 23 | 26 | 28 | 31 | 33 | 36 | 39 | 41 | 44 | 47 |
| 9 | 21 | 24 | 27 | 30 | 33 | 36 | 39 | 42 | 45 | 48 | 51 | 54 |
| 10 | 24 | 27 | 31 | 34 | 37 | 41 | 44 | 48 | 51 | 55 | 58 | 62 |
| 11 | 27 | 31 | 34 | 38 | 42 | 46 | 50 | 54 | 57 | 61 | 65 | 69 |
| 12 | 30 | 34 | 38 | 42 | 47 | 51 | 55 | 60 | 64 | 68 | 72 | 77 |
| 13 | 33 | 37 | 42 | 47 | 51 | 56 | 61 | 65 | 70 | 75 | 80 | 84 |
| 14 | 36 | 41 | 46 | 51 | 56 | 61 | 66 | 71 | 77 | 82 | 87 | 92 |
| 15 | 39 | 44 | 50 | 55 | 61 | 66 | 72 | 77 | 83 | 88 | 94 | 100 |
| 16 | 42 | 48 | 54 | 60 | 65 | 71 | 77 | 83 | 89 | 95 | 101 | 107 |
| 17 | 45 | 51 | 57 | 64 | 70 | 77 | 83 | 89 | 96 | 102 | 109 | 115 |
| 18 | 48 | 55 | 61 | 68 | 75 | 82 | 88 | 95 | 102 | 109 | 116 | 123 |
| 19 | 51 | 58 | 65 | 72 | 80 | 87 | 94 | 101 | 109 | 116 | 123 | 130 |
| 20 | 54 | 62 | 69 | 77 | 84 | 92 | 100 | 107 | 115 | 123 | 130 | 138 |

From Siegel (1956); after Auble (1953).

## A5 Critical values of chi-squared

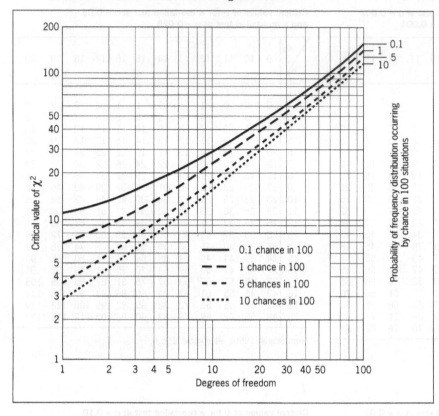

*Source:* based on McCullagh (1974)

# A6 Useful contact details

## Aerial photographs

www.oldaerialphotographs.com
*Historic UK aerial photographs*

## Census and other government data

Census Services
General Register Office for Scotland
Ladywell House
Ladywell Road
Edinburgh
EH12 7TF
www.gro-scotland.gov.uk
*Sales of Scottish census data*

Census Customer Services
Office for National Statistics
Segensworth Road
Titchfield
Hampshire
PO15 5RR
www.ons.gov.uk
*Sales of census data*

www.police.uk
www.crime-statistics.co.uk
*Local crime data*

www.neighbourhood.statistics.gov.uk
*All census and crime data*

## Diseases

www.envhealthatlas.co.uk
*Maps 14 diseases in England and Wales down to ward level*

## The Environment

British Geological Survey
Keyworth
Nottingham
NG12 5GG
www.bgs.ac.uk
*Sells maps and reports on sites of geological and geomorphological interest*

Countryside Council for Wales
Maes y Ffynnon
Penrhosgarnedd
Bangor
Gwynedd
LL57 2DW
www.ccw.gov.uk
*Acts in the same capacity in Wales as Natural England*

Department for Environment, Food
and Rural Affairs
www.defra.gsi.gov.uk
*Flooding, forestry, pollution, food, farming, fisheries, rural communities and conservation*

Environment Agency
www.environment-agency.gov.uk
*Water, rivers, flooding, pollution, waste management – excellent website*

Field Studies Council
Montford Bridge
Shrewsbury
Shropshire
SY4 1HW
www.field-studies-council.org
*Runs field studies centres around the country where GCSE and A-level field courses are offered; publishes a wide range of useful fieldwork publications*

Friends of the Earth
139 Clapham Road
London
SW9 0HP
www.foe.co.uk
*Publishers of many environmental leaflets*

Geographical Association
160 Solly Street
Sheffield
S1 4BF
www.geography.org.uk
*Many publications helpful for students and teachers alike; 'Classic Landforms' series*

The Geological Society
Burlington House
Piccadilly
London
W1V 0JU
www.geolsoc.org.uk
*Publishes information about geological sites*

Greenpeace
www.greenpeace.org.uk
*Publishers of many environmental leaflets*

National Parks
Information on all the UK National
Park websites can be accessed through:
www.nationalparks.gov.uk

Natural England
Foundry House
3 Millsands
Riverside Exchange
Sheffield
S3 8NH
www.naturalengland.org.uk
*Much on the website relating to forestry, landscape, National Parks, Areas of Outstanding Natural Beauty, Heritage Coasts and National Trails*

Royal Geographical Society
1 Kensington Gore
London
SW7 2AR
www.rgs.org
*Expeditions advice, library and map room available to Fellows and Members; Fieldwork ideas and details of fieldwork methods on website*

Scottish Environmental Protection
  Agency
Erskine Court
Castle Business Park
Stirling
FK9 4TR
www.sepa.org.uk
*Exists to protect land, air and water in Scotland*

Scottish National Heritage
www.snh.org.uk
*Publishers of descriptions of nature reserves and other information*

Wildlife Trusts
The Kiln
Mather Road
Newark
Nottinghamshire
NG24 1WT
www.wildlifetrusts.org
*A conservation organisation linking 47 individual Wildlife Trusts*

## Environmental monitoring

Casella Measurement
Regent House
Wolseley Road
Kempston
Bedford
MK42 7JY
www.casellameasurement.com
*Suppliers of meteorological, air and noise measuring equipment*

Geo Supplies Ltd
www.geosupplies.co.uk
*Suppliers of geological and surveying
equipment*

Griffin Education
www.griffin-education.co.uk
*One of the largest UK suppliers of scientific
equipment and chemicals, including
meteorological instruments and equipment
for field surveying*

Hydrographic Office
www.ukho.gov.uk
*Admiralty charts*

Instrument Depot
www.instrumentdepot.com
*Air pollution monitoring including Dräger
tubes*

Meteorological Office
www.metoffice.gov.uk
*Weather data*

National River Flow Archive
Maclean Building
Crowmarsh Gifford
Wallingford
Oxfordshire
OX10 8BB
www.ceh.ac.uk
*Publishers and suppliers of detailed rainfall,
groundwater and river flow data for over
1300 gauging stations.*

Ozone Solutions
www.ozonesolutions.com
*Suppliers of ozone detectors*

Philip Harris
www.philipharris.co.uk
*Major suppliers of instruments to schools:
meteorology, surveying, ecology, soil studies*

## Farming and forestry

Department for Environment, Food
  and Rural Affairs
www.defra.gsi.gov.uk
*Provides published information on a wide
range of farming and fishery topics*

Forestry Commission
www.forestry.gov.uk
*Publications about afforestation and its
impact*

## GIS applications

www.ordnancesurvey.co.uk/oswebsite/
getamap
www.magic.gov.uk
*Environmental data in mapped form*

www.environment-agency.gov.uk
*Interactive mapping tools to examine the
local environment*

www.getmapping.com

www.multimap.com

www.flashearth.com

www.esri.com
*A leading supplier of GIS software with free
ArcGIS trial*

## Maps and satellite images

Environmental Record Search
www.aerialphotos.com
*Historical UK aerial photographs*

Geology Maps
www.bgs.org.uk

Getmapping
www.getmapping.com
*Aerial photographs covering the whole UK*

Goad Maps
www.experian.co.uk
*Suppliers of Goad Plans – shopping-centre
layouts*

Google
www.google.co.uk
*Online aerial images of large parts of the
World*

Natural England
www.naturalengland.org.uk
*Maps and data in the form of GIS datasets
about the natural environment of England*

Land Use Maps
www.geoinformationgroup.co.uk
www.ceh.ac.uk
*Land cover maps provided free to students*

Old OS Maps
www.old-maps.co.uk

Ordnance Survey
www.ordnancesurvey.co.uk
www.ukmapcentre.com
*Contact for sales of OS maps, aerial
photographs and GIS maps*

Soil Maps
www.bgs.ac.uk/nercsoilportal

Stanfords Ltd
12–14 Long Acre
Covent Garden
London
WC2E 9LP
www.stanfords.co.uk
*Map shop, supplying a wide range of maps
including geological maps*

## Recreation and tourism

www.visitengland.org
www.visitbritain.org
www.visitscotland.org
*Excellent data sources*

## Urban Geography

Big cities
Geographical Association
Discovering Cities guides
www.geography.org.uk

Property Prices
www.zoopla.co.uk
www.voa.gov.uk
*For business rateable values and house
council tax evaluations*
www.houseprices.landregistry.gov.uk

Historic Business Directories
www.kellysdirectories.com
www.le.ac.uk

# A7 Sources and further reading

## General texts and websites

Clifford, N., French, S. and Valentine, G. (2010), *Key methods in geography*, Sage

Crampton, J. (2010), *Mapping: A Critical Introduction to Cartography and GIS*, Wiley-Blackwell

DeMers, M. (2009), *GIS for Dummies*, Wiley

Field Studies Council (FSC) website *(geography-fieldwork.org)*

Geographical Association website *(www.geography.org.uk)*

Hammond, R. and McCullagh, P. (1978), *Quantitative Techniques in Geography*, Oxford University Press

Heywood, I., Cornelius, S. and Carver, S. (2011), *An Introduction to Geographical Information Systems*, Prentice Hall

Holmes, D. and Farbrother, D. (2000), *A-Z Advancing Geography: Fieldwork*, Geographical Association

May, S. and Richardson, P. (2005), *Fieldwork File: Managing Safe and Successful Fieldwork*, Geographical Association

Royal Geographical Society website: fieldwork techniques *(www.rgs.org)*

Widdowson, J. and Parkinson, A. (2013), *Fieldwork Through Enquiry*, Geographical Association

## Physical and environmental studies

Atkinson, B. W. (1985), Update: *The Urban Atmosphere*, Cambridge University Press

Barry, R. and Chorley, R. (2009), *Atmosphere, Weather and Climate*, Routledge

Cooke, R. U. and Doornkamp, J. C. (1990), *Geomorphology in Environmental Management*, Oxford University Press

Dunlop, S. (2014), *Meteorology Manual: The practical guide to the weather*, Haynes

Edmondson & Roberts (1997), *Plants common on sand dunes*, Field Studies Council

Environment and Health Atlas for England and Wales (2014) *(www.envhealthatlas.co.uk)*

Goudie, A. (1990), *The Landforms of England and Wales*, Blackwell

Goudie, A. and Gardner, R. (1985), *Discovering Landscape in England and Wales*, Routledge

Goudie, A. et al. (1990, 2nd edn), *Geomorphological Techniques*, Routledge

Gregory, K. J. and Walling, D. E. (1973), *Drainage Basin Form and Process*, Edward Arnold

Hansom, J. (1988), *Coasts*, Cambridge University Press

Holden, J. (2012), *An introduction to physical geography and the environment*, Pearson

Holmes and Warn, S. (2010), *Understanding geography fieldwork 3: The river environment*, Field Studies Council

Moncrieff, D. and Norman, S. (2010), *Sand dunes: a practical fieldwork manual*, Field Studies Council

Oke, T. R. (1987, 2nd edn), *Boundary Layer Climates*, Routledge

Oldham and Roberts (1999), *Saltmarsh plants of Britain*, Field Studies Council

Orton, Haines and Proctor (1995), *Freshwater investigations*, Field Studies Council

Richardson, D. H. S. (1992), *Pollution Monitoring with Lichens*, Richmond Publishing

Trudgill, S. (1989), *Soil types: a field identification guide*, Field Studies Council

Ward, R. and Robinson, M. (1999), *Principles of Hydrology*, McGraw-Hill

## Urban and rural studies

Daniels, P., Sidaway, J., Bradshaw, M. and Shaw, D. (2012) *An introduction to human geography*, Pearson

Flowerdew, R. and Martin, D. (2005), *Methods in Human Geography*, Pearson

Geographical Association *Discovering cities* series

Gould, P. and White, R. (1986, 2nd edn), *Mental Maps*, Allen and Unwin

Hall, T. and Barrett, H. (2011), *Urban Geography*, Routledge

Knox, P. and Pinch, S. (2006) *Urban Social Geography: an introduction*, Prentice Hall

Morgan, M. A. (1979), *Historical Sources in Geography*, Butterworths

Smith, J. and Yates, E. (1968) *On the dating of English houses from external evidence*, Field Studies Council

Warn, S. and Holmes (2007) *Understanding Geography Fieldwork 2: The Central Business District*, Field Studies Council

Woods, M. (2004), *Rural Geography: Processes, Responses and Experiences in Rural Restructuring*, Sage

# Index